The Deregulation of Natural Gas

A Conference Sponsored by the
American Enterprise Institute for Public Policy Research

The Deregulation of Natural Gas

Edited by Edward J. Mitchell

American Enterprise Institute for Public Policy Research
Washington and London

346.7304
D44
125717
aug. 1983

Library of Congress Cataloging in Publication Data
Main entry under title:

The Deregulation of natural gas.

(AEI symposia; 83B)
1. Gas, Natural—Law and legislation—United States—
Congresses.   2. Gas, Natural—Government policy—United
States—Congresses.   I. Mitchell, Edward John,
1937–   .   II. American Enterprise Institute for
Public Policy Research.   III. Series.
KF1870.A75D47   1983   346.7304'68233   83-9989
ISBN 0-8447-2246-4   347.306468233
ISBN 0-8447-2245-6 (pbk.)

AEI Symposia 83B

1 3 5 7 9 10 8 6 4 2

Printed in the United States of America

Contributors

Catherine Good Abbott
Department of Energy
(now with the Interstate Natural Gas Association of America)

Danny Boggs
The White House

Michael Canes
American Petroleum Institute

Mark Cooper
Consumer Energy Council

Jack Earnest
Texas Eastern Pipeline

Ted Eck
Standard Oil of Indiana

Edward W. Erickson
North Carolina State University

Gordon Gooch
Baker and Botts

Edward Grenier
Sutherland, Asbill and Brennan

George Hall
Charles River Associates

Henry D. Jacoby
Massachusetts Institute of Technology

Robert Leone
Harvard Business School

Robert C. Means
Federal Energy Regulatory Commission

Edward J. Mitchell
American Enterprise Institute

William Niskanen
Council of Economic Advisers

Milton Russell
Resources for the Future

Benjamin Schlesinger
American Gas Association
(now with Booz, Allen and Hamilton, Inc.)

Philip Sharp
United States House of Representatives

Charles Stalon
Illinois Public Utility Commission

William Stitt
ICF, Inc.

Grant Thompson
Conservation Foundation

Stephen A. Watson
Department of Energy

Robert Woody
Lane and Mittendorf

Arthur W. Wright
University of Connecticut

Benjamin Zycher
Council of Economic Advisers

*This conference was held at
the American Enterprise Institute for Public Policy Research
in Washington, D.C., on March 11–12, 1982*

Contents

Preface
Edward J. Mitchell

PART ONE

Overview of Policy Issues: A Preliminary Assessment   3
Milton Russell
Commentary
Edward W. Erickson   33
Benjamin Schlesinger   36
Discussion   40

PART TWO

Pitfalls on the Road to Decontrol: Lessons from the Natural Gas
Policy Act of 1978   53
Catherine Good Abbott and Stephen A. Watson
The Intrastate Pipelines and the Natural Gas Policy Act   71
Robert C. Means
Commentary
Jack Earnest   108
Robert Leone   109
Discussion   112

PART THREE

The Gordian Knot of Natural Gas Prices   125
Henry D. Jacoby and Arthur W. Wright
Commentary
Michael Canes   149
Mark Cooper   151
Discussion   154

Preface

Few public policy issues are as complex and controversial as the deregulation of natural gas. Natural gas legislation, court cases, and state and federal regulatory commissions have produced a labyrinth of regulations that only the determined should approach. In addition to this complexity and confusion, regulation of natural gas markets has caused shortages and price and supply disparities. An indication of the controversy involved is the length of the legislative battle that produced the Natural Gas Policy Act of 1978 (NGPA); Congress debated for seventy-five tumultuous weeks before it passed a natural gas bill.

In March 1982 AEI brought together leading academicians, industry representatives, and government officials to discuss the effect of the NGPA on energy markets and to appraise policy alternatives and their likely outcomes. The papers in this volume represent the state of the debate at the time of the conference. And, although events have proceeded apace since the conference—gas prices rising to what many believe are market equilibrium levels, oil prices continuing their decline, and new policy issues emerging— the conference papers provide background information and analyses that will make a contribution to the legislative debate over deregulation.

Milton Russell opened the conference with a conceptual analysis of price formation under the NGPA and compared the results of the NGPA with those of a deregulated market. He also identified and discussed policy issues, including the potential for large price increases in 1985, the effect of contract provisions on gas prices, and the effects of partial as compared with complete decontrol on economic efficiency and on the distribution of income among gas industry participants. The second session focused on "market-ordering" issues. Robert Means of the Federal Energy Regulatory Commission analyzed the potential problem of the different capacities of interstate and intrastate pipelines (the dual market) to bid for "new" gas, given the differential effect of the NGPA price ceilings and deregulation schedule. Catherine Abbott, then of the Department of Energy, analyzed two market-ordering issues: the "price cushion" (the uneven distribution of low-cost gas) and the effect of contracts on gas prices. The conference concluded with a paper by Dr. Henry Jacoby and Dr. Arthur Wright, of the Massachusetts Institute of Technology and the University of Connecticut, respectively. They suggested the metaphor of the mythical Gordian knot as a way to understand the dilemmas of natural gas policy making. After viewing the history of natural gas regulation, they discussed the advantages and disadvantages of four broad policy options, including immediate decontrol and letting the NGPA run its course.

EDWARD J. MITCHELL

Part
One

Overview of Policy Issues:
A Preliminary Assessment

Milton Russell

This paper sorts through the issues that have been raised regarding current and prospective natural gas field-price regulation and advances some tentative conclusions about what factors should be at the center of the policy debate. Where possible, preliminary judgments are made about how the weight of the evidence and the analysis come out on these matters.

There are a number of policy alternatives that could dictate the choice of issues and factors to be considered—ranging from immediate and full decontrol to permanent extension of controls. The alternative around which this paper is organized, however, is full implementation of the Natural Gas Policy Act of 1978 (NGPA).[1] From this starting point outcomes are projected, alternatives defined, and consequences analyzed.

This paper presents a highly compressed, selective, and simplified analysis of alternatives for natural gas control. It is designed to give the nonspecialist a starting point from which understanding of the issues can begin. To serve this purpose the presentation must necessarily proceed from point to point, but it is the essence of the problem that all these issues must be considered simultaneously and in a general equilibrium framework. The temptation to interrupt the flow constantly with caveats and asides has been resisted in the belief that to do otherwise would risk getting some points more nearly right only at the expense of making the whole more confused.[2]

This paper incorporates revisions, made immediately after the conference, based on comments by Benjamin Schlesinger, Edward W. Erickson, and other participants. No further revisions were made to take account of subsequent developments or later research.

W. David Montgomery, Thomas J. Lareau, and Mark Lyons made helpful suggestions on the original draft, as did Michael Coda and Sandra Glatt, all of Resources for the Future; the last two also provided research assistance. The author wishes to express appreciation to the numerous researchers and persons concerned with natural gas policy who gave generously of their time and provided unpublished materials useful in preparing this paper. The views and conclusions expressed here are those of the author and should not be attributed to the reviewers, conference participants, or colleagues at Resources for the Future.

This paper starts with three observations that are peripheral to its major thrust but may assist the reader in assessing the consequences of adopting different policy alternatives. The next section of the paper develops a framework for thinking about natural gas regulation by describing a simple model of price formation under the NGPA. This model suggests a number of issues or problems that will attract the attention of those who must come to some judgment on how the gas market should be treated. Each of these is then discussed in turn.

Three Observations

The first observation is that almost without regard to what happens with regulation, massive changes in natural gas field markets are under way that will fundamentally transform the natural gas industry. These changes will withdraw two related foundation blocks on which this large and pervasive industry has been built for the past three decades and more. The first is "cheap" gas, whether caused by economic-technological-institutional factors (as it was through the early 1960s) or by regulation. At the going price there has usually been excess demand for gas, and, except for special circumstances, finding markets has been no problem. The second foundation block is field-price regulation itself, and the economic, regulatory, and political baggage it has carried with it. Regulation produced expectations about gas prices and availability of supply that affected the location of consuming industries, the criteria by which fuel choices were made by consumers, growth strategies and financial structures of pipelines and distributors, terms of contracts at every stage of the gas industry, energy research and development programs, and political coalitions. It affected the kinds of management style that prospered, the sorts of investors that were attracted, and the state and local regulatory scrutiny that resulted.

The twin building blocks of cheap gas and a regulated environment cannot be removed without far-reaching effects, only some of which are discussed in this paper. The operations and finance of pipelines and distributors will be altered, and consumers' choices will change. Opportunities for new enterprises will arise and old functions wither; remaining regulatory responsibilities will shift. Those in the industry who are not positioning themselves to respond to this transformation may be in serious jeopardy.

The second observation has to do with the tendency of industry analysts to concentrate on either form or substance. By substance, what is meant is the underlying economic forces that are driven by the self-interest of buyers, sellers, and voters; by form, existing regulations, contract clauses, and customs of the industry. As to the latter, when fundamental economic shifts of the kind now gripping this industry occur, the institutional trappings through which the industry is managed and organized will be changed or swept away,

4

as the experience of the international oil industry after 1970 demonstrates. Hence it is misleading if the future is projected on the basis of steadfast adherence to what contracts say, how the industry is now organized, and what regulations exist and now mean.

At the same time, textbook economic analysis also will not be a precise predictor of developments, certainly not in the short run. Whatever market changes take place, such factors as indefinite pricing clauses, purchased-gas adjustment regulations, take-or-pay provisions, and pipelines' obligations to serve cannot be cast aside immediately and will affect the outcomes. As a generalization, however, it does seem to me that the regulatory milieu in which most people in this industry have operated has led them to unwarranted reliance on what are ultimately fragile institutions in judging how matters will actually work themselves out.

The third observation is that the lags between inducements to act and results in the gas industry can be long. Institutional drags and technological realities are both involved. Lags have two effects. First, it takes a long time for changes in economic conditions that are "surprises" to have an effect on outcomes, and thus shocks are in some ways dampened as they move through the system. On the supply side, when incentives are increased, it takes several years for the full effects to be felt, that is, for resources to be deployed or redeployed to more profitable prospects and for gas to come on the market. Lags occur when incentives decline as well—expenditures already made often dictate bringing prospects to completion even if it is obvious midway through that total returns will be negative. On the demand side, some lags are substantially shorter. As noted below, dual fuel capacity exists in major fuel-using installations, and substantial changes in gas use can be accomplished with the turn of a valve. When equipment must be installed or modified, however, consumption lags can also be long.

The second effect of lags is that expectations play an enormous role in the behavior of the industry. Things start happening long before new regulations or laws become effective, and analysts and legislators need to bear this in mind. An industry whose investments come to fruition only three to ten years in the future and last a generation must be looking far ahead; if the picture is uncertain, the tendency is to avoid commitments. The important role expectations play presents both an opportunity and a challenge to policy. The opportunity is to bring about change with less disruption by announcing policy shifts with sufficient notice that adjustments can occur in an orderly way. The challenge is to avoid debilitating uncertainty by making policies as clear as possible. The difficulty here is that the political and industry timetables are often not synchronized. Political action usually occurs only when issues are "ripe"—when the need for action is urgent and at the focus of public attention. By that time it is often too late for efficient response by the industry.

5

A Simple Model of Prices under the Natural Gas Policy Act

The entry point for an analysis of the issues surrounding actual gas policy is through an examination of processes by which prices will be formed under its application. Once these processes are understood and some projections are made, the incentives for altering or sustaining the NGPA can be illuminated and the mechanisms for making changes indicated.

Three distinct gas price outcomes are crucial to policy. The first is the path of prices under the NGPA before partial decontrol in 1985, the second is the average wellhead price in 1985 and beyond, and the third is the pattern of prices both before and after partial decontrol. What is important, of course, is how these outcomes compare with those under other decontrol alternatives, specifically, full decontrol imposed at the same time that partial decontrol would occur under the NGPA and gradual full decontrol culminating in 1985. (The option of reimposition of controls is outside this discussion.) Once these outcomes are understood, their implications can be assessed, and policy decisions can be clarified. A simple model of gas price formation under the NGPA can assist in this effort.

Fuels Competition at the Burner Tip. Natural gas is used for many purposes, and in some of them consumers would be willing to pay a great deal for continued supplies. Even in these uses, however, consumers have some opportunity to economize as the price rises. A significant proportion of gas is not used for residential and commercial purposes or even for feedstock or special-process heat applications in industry, where its characteristics of cleanliness and ease of control are important. About 30 percent of the gas goes under boilers as a source of bulk heat—a function that can be performed by heavy fuel oil or in some cases by coal or nuclear power.[3] The supply of gas is sufficiently large that when all higher-valued uses are satisfied, a significant quantity will remain to be sold as a source of bulk heat. Obviously, therefore, it must compete with alternative fuels on the basis of price at the bulk heat burner tip, with the price per delivered Btu adjusted for relative capital cost of the equipment, burning efficiency, cost of combustion and cleanup, convenience, and certainty of supply. Going further, to expand quickly into new markets, gas must be priced sufficiently below the competitive fuel to offset the capital cost and inconvenience associated with switching fuels.

The price of coal per delivered Btu (adjusted as described above) is, and has been anticipated to be, below the comparable gas and oil prices in most locations and applications. Thus, where coal is an otherwise acceptable choice over oil, it is already being consumed, or sufficient incentives already exist that conversion to it will occur over time. Given reasonable expectations about oil prices (set on the world market) and availability, it appears that a significant amount of oil will continue to be used for bulk heat and that gas will share this market. Thus, the price of heavy fuel oil (as adjusted) in the long run will probably be the dominant influence on the unregulated natural gas price at the burner tip in bulk heat uses. This does not mean, of course, that the price of gas per Btu will be exactly equal to the price of oil, even when adjusted for relative sulfur content, burning efficiency, and costs; because gas and oil are not perfect substitutes even in bulk heat markets, the gas supply will have some effect on price. (There is some chance that the supply of gas would be sufficiently large, and the industrial demand sufficiently low, that oil would be backed out of this market and gas at the margin would compete with coal. This case is not considered.)

Questions remain about how this bulk heat price at the burner tip affects prices for gas in other uses and at the field-market level, and how much time is required for gas prices to adjust to changes in oil prices. Here the issue becomes complicated because it involves the speed with which major changes in gas use can take place, transportation costs, and rate structures of pipelines and distributors, which are affected by federal and state regulation.

Price Adjustment at the Burner Tip. Adjustments of gas use to changes in oil prices depend in the short run on the amount of dual-fired capacity that exists. The existing data and evidence are distorted by field-price regulation, by the incremental pricing provision of title II of the NGPA, by the "off-gas" provisions of the Power Plant and Industrial Fuel Use Act, and by consumers' uncertainty about the reliability of gas supply.[4] It appears, however, that in 1981 about 10 percent of the total gas supply was used in installations where oil could be substituted and that oil was used in installations where about 5 percent of the available gas could be used if it were available at a competitive price.[5] It is not clear, however, whether delivery constraints or other factors might limit ready substitutability to a smaller proportion of the market than these numbers imply. Looking to the future, inherent uncertainty about relative oil and gas prices and availability suggests that major fuel users will continue to be willing to bear added investment and other costs to maintain the option to switch among fuels.

This analysis suggests that, overall, enough flexibility will exist in the end-use markets to prevent substantial and sustained differences between the burner tip price of gas and the adjusted price of heavy fuel oil in the absence of regulation of various sorts. Even so, transitory and localized differences can exist. For a simple model of gas prices under effective decontrol, however, it seems a reasonable first approximation to assume that gas prices will adjust to oil prices relatively completely and quickly. Market shares will change to bring gas prices back in line when there are initial shifts in the relative prices of oil and gas—unless these price changes are very large and sudden.

7

From Burner Tip to Wellhead Price: Transportation Costs. The price of fuel oil is not heavily influenced by location; fuel oil is cheap to transport per unit of heat value (at least by water, and major use points are accessible to barges), and refineries are widely dispersed. Thus the oil price with which gas will compete will be not very dissimilar around the nation. In contrast, natural gas deposits that provide the incremental supply and therefore are significant in setting overall prices are concentrated in the Southwest and along the Gulf Coast, and gas is costly to move to market.

Suppose the producing-region bulk heat markets could absorb all the gas available and still not totally eliminate some fuel oil use by firms that would be willing to switch to gas if the price were slightly lower. In that case the price at the wellhead would be the adjusted fuel oil price at the burner tip in the producing region, less small transport costs. If the local bulk heat market were effectively swamped with gas, however, the wellhead price would need to be lower to sell all gas produced—so as to allow for the extra transport cost to more distant markets. This artificial market separation by location is too simple, of course; in actual market situations each potential purchaser has a different price at which different quantities of oil and gas would be used.

Although uncertainty exists, the weight of the evidence suggests that the quantities of gas that would be available are such that the bulk fuel market would be penetrated (though to different degrees) in regions distant from the producing region. At the same time, the total market shared between gas and fuel oil is sufficiently large that gas supplies would not be adequate to displace oil fully. Consequently, the overall demand for gas in a regime of effective decontrol would be relatively elastic. The share of the burner tip market captured by gas would depend on the quantity supplied at the price that could penetrate bulk heat markets distant from the producing regions. Hence the maximum average price in the field markets would be somewhat *lower* than the adjusted price of competing fuel oil *in* the gas-producing region when taking only transport costs into account.

From Burner Tip to Wellhead Price: Pipeline and Distributor Charges. Different types of consumers of gas require different pipeline and distributor expenditures to serve them. Small users and those with highly seasonal loads, for example, cost more to serve than large, steady customers. Aside from these cost differences, however, different consumers also respond differently to higher prices—the consumption by small, seasonal customers is generally much less responsive to changes in price (less elastic) than is that of large industrial consumers. The latter have more opportunities to switch fuels, and for them the ratio of fuel cost to capital cost is generally higher because fuel use is not seasonal. These differences in demand elasticity offer pipelines and distributors an opportunity and an incentive to shift the burden of general system costs, including total gas purchase costs, among different consumer classes.

Within certain market and regulatory constraints, the opportunities for profit of pipelines and distributors are generally enhanced with added flow of gas through the system. Consequently, there is often an incentive for them to hold the price down to the customers with the most elastic demand—so as to compete effectively for their business—while covering costs by increasing prices to those customers with the most inelastic demand. In this way less volume will be lost per unit of extra revenue obtained. Regulatory authorities, while maintaining a healthy distribution and pipeline system, must balance conflicting goals: a desire to be "fair" overall; a desire to "protect" the most defenseless of the customers from high prices, whether "fair" or not; and a desire to promote economic development in the state or locality by providing cheaper gas service to the industrial customers with the highest demand elasticity. Regulatory authorities must also be conscious of what is happening elsewhere—with regard to gas prices paid by industrial consumers in other jurisdictions, for example. They will be sensitive to competitive advantages obtained or lost by employers in their jurisdictions.

The implication of this pricing discretion on the part of actors downstream from a wellhead is clear: the average field price arises from the subtraction of gas transportation and delivery *charges*—not costs—from the heavy fuel oil price at the burner tip for the marginal bulk heat consumer. To the extent that those charges are lowered for industrial customers, the market-clearing field price will be higher, the share of the bulk heat market served by gas greater, the quantity of heavy fuel oil used smaller, the quantity of gas produced greater, and, for non–bulk heat customers, the price of gas higher and consumption lower. As noted below, this pricing discretion is also important because it will affect the situation of consumers differently depending on the size of the "old-gas cushion" of the pipelines from which they obtain gas.

The Burner Tip Price of Fuel Oil. The process by which gas prices at the burner tip are transmitted to the field price has been outlined, and the tendency of the burner tip price to flow from and to approximate the heavy fuel oil price paid by the marginal consumer has been explained. But projections of gas prices require some estimate of what that heavy fuel oil price will be. What clues can analysis provide?

Fuel oil costs will be determined by a number of factors, certainly including the price of crude oil, but no simple rule translates a projected "marker" crude oil price into a heavy fuel oil price to bulk heat users. Other factors are involved: the mixture of crude oils entering the refining stream by type and price; the configuration of refineries in their ability at different costs to produce different products; and the relative demand elasticities and therefore profitabil-

9

ity of producing different petroleum products. It is unnecessary to go into these questions here except to note that (1) the overall price of crude oil is by far the most important of these factors; (2) the crude oil stream is likely to be such that heavy fuel oil will be a greater proportion of the total (and therefore cheaper in relation to crude oil) in the future; but (3) this change will probably be more than offset by the ongoing effort to upgrade refineries to produce less fuel oil from each barrel of crude. Although prices will fluctuate with overall oil market conditions and the quantities of fuel oil demanded, it seems reasonable to conclude that prices of fuel oil in relation to crude oil will increase.

Given this relationship between natural gas, fuel oil, and crude oil, what can be said about the prospective market-clearing price for natural gas at the wellhead, taking everything into account? Nothing very precise in quantitative terms, obviously. Studies of the question have tended to conclude, however, that the market-clearing field price per Btu would be about 70 percent of the crude oil price.[6] A price at that level or somewhat higher seems consistent with the analysis given above and with such evidence as is available, but only so long as the crude oil price does not fluctuate wildly or move substantially outside its current range.

Conclusions on the Simple Model. A simple static model of gas field-price formation can now be laid out, keeping in mind the complexities that are described above. The average field price of gas that purchasers can afford to pay will be determined by what incremental units of gas can be sold for in the bulk heat market, minus transport and handling charges. Depending on assumptions about the treatment of various categories of gas and developments between now and 1985, from 40 to 60 percent of all gas will be free of controls at the end of 1984.[7] Though subject to numerous caveats and adjustments (many of which are noted below), this partial decontrol *can* offer sufficient freedom of action within the natural gas industry to result in an *average* price of gas in the field markets not greatly different from that which would come with full decontrol. Many questions exist about whether this outcome *will* occur, and much of the remainder of this paper is devoted to exploring them.

The simple model arrives at the conclusion that average prices under partial and full decontrol will be essentially the same by examining the conditions under which burner tip and field markets will clear. The burner tip market will clear at the adjusted price of heavy fuel oil. This price, in turn, is consistent with an average gas field-market price that differs from it essentially by the amount of transport and handling charges to the marginal consumer. Gas purchasers will be willing to pay on average this price but no more; gas sellers need take no less. With a substantial amount of gas free to seek its own price, this average field price will be formed from two components: the regulated quantities of gas at the controlled price and the unregulated quantities of gas at a price sufficiently high to bring the average price up to the market-clearing level. In this framework each pipeline, whatever its situation, must pay the same price for unregulated gas of similar quality (in many dimensions) because in a free market there is no reason for bargains to be struck at any other price. From this simple model four important implications flow:

1. The basic price to consumers as a group is not much affected by the continued controls on 40–60 percent of the total gas supply.

2. Weighted by volume, the average price of unregulated gas is held above the market-clearing level about to the extent that the regulated price is held below, and producers as a group are not much affected by continued partial regulation as compared with full decontrol.

3. Therefore, as a first approximation, the NGPA as compared with complete decontrol shifts revenues among producers, not between producers and consumers.

4. Because of the different patterns of gas produced and used under the NGPA from those under full decontrol, the relative efficiency of resource use under the two regimes is a matter of policy concern.

A Caution about This Analysis. One implicit assumption behind this view of how the gas market works needs to be made clear because it may significantly affect the outcome. This model works *as if* the newly deregulated and unregulated gas were sold on a spot or very short-term contract market, while in fact much of it will be sold under contracts of varying length. This spot market assumption dictates that unregulated and deregulated prices will be high immediately upon decontrol and will gradually decline as less and less of the regulated-gas cushion remains. It is also assumed that the unsatisfied demand at the burner tip will be large enough to drive prices immediately to the equilibrium market-clearing level.

To what extent is the spot market assumption counterfactual? What difference does it make? The assumption may not be as counterfactual as it appears. First, there is some flexibility in purchasers' behavior. Take-or-pay provisions do restrict their behavior, but the quantities taken are not rigidly fixed. When quantities can vary under contracts with different prices, in some ways the effects are the same as those of a spot market. Second, contract terms for new sales are growing shorter, and the renegotiation clauses in old contracts mean that new, shorter terms can be written into them as well. Third, many recent contracts have "market kick-out" clauses, which allow purchasers to abandon contracts when the terms prove onerous. Finally, with the arrival of partial decontrol, the industry structure may quickly switch toward creation of a short-term contract segment. Already sales among pipelines and short-term sales are becoming more common.

11

The spot market assumption probably makes little difference to the overall conclusions of the model, but might affect the magnitudes involved. In general, under partial decontrol the contract price for uncontrolled gas would be lower initially than the spot price required to clear the short-term market. Buyers would be unwilling to make long-term contracts at a price that would ultimately be above market levels. In parallel, however, shortages may not occur because of lags in consumption increases at the burner tip.

Questions for Analysis. The major policy and analytical issues surrounding the NGPA and its alternatives flow from this simple model and can be interpreted in its framework. These issues pose policy problems that can be summarized in a series of questions:

1. Will the market-clearing price at the time of partial decontrol under the NGPA be substantially above the average field price in late 1984, so that a major price shock will be imposed on the economy?

2. Will the indefinite pricing clauses in gas sales contracts drive the average price above the market-clearing price (under either partial or full decontrol), and if so, how soon, through what mechanisms, and with what ramifications will the average price revert to the market-clearing level?

3. Will the uneven endowment of regulated (low-priced) gas leave some pipelines (particularly intrastate pipelines) and their customers at an unfair disadvantage under partial decontrol and create market disorder—characterized by differential availability of gas to industrial consumers, by differential profitability of pipelines and distributors, and by differential prices to non-industrial consumers?

4. Does the different pattern of field prices under partial rather than full decontrol have a significant effect on the quantity of gas that will be brought to market?

5. What will be the effects of partial as opposed to full decontrol on the distribution of income among gas industry participants?

6. Do the pattern of prices and the behavior of gas producers and consumers under partial as compared with complete decontrol have significant effects on the efficiency with which the economy uses resources to produce goods and services?

7. How does phasing in full decontrol between now and 1985 compare with either partial decontrol under the NGPA or full decontrol at that time?

Answers to these questions can be approached by analysis but must ultimately depend on facts over which there is uncertainty and on events that cannot be predicted. Much research has been undertaken, and many position papers have been produced, however, and from this work some preliminary conclusions can be drawn.

The Price Shock Problem

Will the market-clearing price at the time of partial decontrol under
the NGPA be substantially above the average field price in late 1984,
so that a major price shock will be imposed on the economy?

The field price of natural gas has risen rapidly since the early 1970s, more than
tripling in real terms between 1970 and 1978 and rising 40 percent from the
passage of the NGPA to the end of 1981.[8] The increases in average price under
the regime established by the NGPA have been larger than anticipated. The
result has been to rectify to some degree the failure of the drafters of the NGPA
to anticipate the oil price increase of 1979–1980. The sources of average price
escalation have been documented elsewhere; beyond the real price increases
mandated for some categories of gas, they include the admixture of unexpectedly large quantities of unexpectedly high-priced unregulated deep gas and the
reclassification of flowing gas into higher-priced categories within—and in
some cases undoubtedly outside—the rules Congress established. These developments are not independent of the added incentive to use gas presented by the
increase in the price of the competing fuel; the enhanced ability to market high-
priced supplements to flowing gas supplies has been a motive force behind the
rising prices.

To establish some comparisons, the wholesale heavy fuel oil price to
which the market-clearing natural gas price is keyed rose in real terms from a
little over $3.00 per million Btu at the implementation of the NGPA to about
$4.50 at the beginning of 1982. At its peak in early 1981 it was about $6.00.
The average field-market gas price was about $1.21 per million Btu in early
1979 when the NGPA was implemented and about $2.10 in early 1982. The
gas price per Btu was a little over 40 percent of the crude oil price when the
NGPA passed and is a little less than that now. But in the interim it fell to less
than 30 percent of the crude oil price, and it is now climbing.[9]

The questions are whether gas prices will continue to increase at their
recent rates and what will happen to oil prices, recognizing that the two are not
independent because the path of oil prices will influence how fast gas prices
increase, even under the NGPA.

My judgment is that the price of oil by 1985 is unlikely to be substantially
different in real terms from what it is now, with a decline in the interim. At the
1978–1981 rate of increase in gas prices, this would bring the natural gas price
to a little more than 60 percent of that of crude oil, or not far from what is
assumed to be a market-clearing level. Moreover, because of the lags built into
price adjustments by staggered contract reopening dates and because expectations of a price jump will bring anticipatory action, the rate of the transition
will be slowed.

Nonetheless, the prospect of a price jump remains even if current trends are maintained. Moreover, there are a number of reasons to believe that the rate of gas price escalation since 1978 will not be sustained and hence that a significant gap will exist between average prices in late 1984 and the market-clearing average price in 1985.[10] For one thing, pipelines and other purchasers of gas in the field will become more cautious in bidding very high prices for uncontrolled gas, for a number of reasons. First, oil prices are now not expected to rise as fast as they were previously, and consequently expectations are that higher-priced gas will find less of a market than was earlier thought. Second, a shift downward in expectations about the long-term natural gas price will cool the speculative fever that previously motivated long-term purchases at currently noncompensatory costs. Pipelines that thought they were in equilibrium in the long term may find that their projected gas costs are now too high, leading to a pullback from the market. Third, as partial decontrol approaches, large quantities of uncontrolled gas will reach the market. There will be less low-cost gas to cushion each unit of high-cost gas purchased. Fourth, as the field price of gas rises toward the market-clearing level, there will be correspondingly less room to pay high prices and still market gas.

The most profitable opportunities to reclassify gas into higher categories were obviously taken first. Additional steps will be less productive as the range of possibilities narrows toward those in which substantial investment may be required.

The overall supply and demand for gas have changed, which may also dampen the rise in gas prices. Gas supplies have held up well despite relatively conservative drilling growth, and the response to already higher prices has led to lower sales or less unsatisfied demand than anticipated. The data are flawed, of course, on both sides: some of the gas finds have been associated with booming oil exploration; declines in demand have resulted from the recession and slow growth, from uncertainty about the prospects of gas supplies, and from legislated efforts to restrict gas use. The size of the "hidden" gas demand in the industrial market is unknown.

Finally, prospective gas purchasers will become more cautious as uncertainty arises about the future operation and makeup of the industry. As suggested earlier, decontrol under the NGPA or otherwise will bring substantial strains to the downstream sectors of the gas industry, sectors that are already under stress because of high interest rates. Added financial exposure through connecting additional gas supplies, especially if high prices and onerous take-or-pay provisions are involved, may appear excessively risky, although in a different environment they would appear quite acceptable. Purchasers may become content to tolerate lower deliverability cushions until uncertainty diminishes.

Another factor works in the opposite direction—toward faster average price increases between now and 1985 and thus a smaller shock then. Because of lags between changed incentives and investment and between investment and output, provisions of the NGPA that allow uncontrolled or higher regulated prices for some gas have not yet been fully reflected in gas sales. Additional high-priced volumes will be coming on stream and may contribute more to future price escalation than gas of similar status has over the past two years.

One further source of price shock uncertainty should be mentioned. The somewhat flexible connection between reserves secured and deliveries required means that expectations play a major role in field-price behavior. Consequently, substantial volatility is possible if oil prices move upward quickly or expectations turn around on some other front. Most possible surprises would drive the gas price upward (that is, even ignoring the "contracts" problem discussed in the next section).

Quantitative estimates of the size of the price shock are speculative. Weighing the factors discussed above, however, it is reasonable to conclude that under the NGPA there will remain an opportunity for a significant price increase in 1985—based on market forces alone.

The "Contracts" Problem

Will the indefinite pricing clauses in gas sales contracts drive the average price above the market-clearing price (under either partial or full decontrol), and if so, how soon, through what mechanisms, and with what ramifications will the average price revert to the market-clearing level?

Numerous natural gas contracts have contingency clauses of different sorts that define the way prices will be determined in the event of decontrol. The pricing provisions reflect parties' expectations and their respective bargaining power and were agreed to as part of a bargaining process that included numerous other provisions of the contract, some of which undoubtedly would have been affected had the pricing provision been different.

The "contracts" problem arises because some contracts have provisions that would drive their prices above both long-term and partial decontrol market-clearing prices.[11] These contracts call for a price to be established as a percentage of the price of some petroleum product, often 110 percent of No. 2 fuel oil. Other contracts contain "most-favored-nation" clauses, which hold that the price will be based on some set of the highest prices paid by or to others.

A number of effects are associated with contract clauses that put some prices above the market-clearing level. Producers as a group could be enriched (at least temporarily) at the expense of consumers *and* of pipelines and distributors, whose returns would be squeezed. Some producers would be benefited at the expense of others without such contracts, who could obtain less for their gas because the higher prices paid for some gas would depress the price other gas could command. A further concern is that the supra-market-clearing prices

15

would not be distributed equally among pipelines and that this would cause regional and other disparities. This issue is considered in another context below.

Contracts of the form described undoubtedly exist, and their operation if not modified would have the consequences specified. These consequences would follow under either partial or complete decontrol, although the size of the benefits and costs borne by different parties would vary with the proportion of contracts decontrolled. The questions are the size of the potential problem and whether existing mechanisms are adequate to moderate the more serious consequences of the disorder that would result.

Clearly, in the long run, contracts not in the interest of the parties will not be honored. As Arlon R. Tussing and Connie C. Barlow put it, "Contracts will not determine the price of gas. The market value of gas, rather, will determine which contracts will be honored, which will be renegotiated, and which will be repudiated with impunity."[12] For this process to work without massive and disruptive litigation requires, however, that contracts be renegotiated by parties concerned (selfishly, to be sure) with each other's long-term interest. Leaders of the industry on both sides of the market will concur that reasonable parties could agree to renegotiate improvident contracts to an acceptable solution. The question is, Given the dynamics of the decision process, will it happen?

Probably not. Whatever the statesmanlike posture of corporate leaders, actual field-by-field, sale-by-sale decisions will be made by persons far down the corporate ladder. They will want to impress on their superiors that *they* are able to drive the best bargain possible. No gas sales manager of a major company will want to be the first to relinquish a price firmly tied up in an existing contract. Smaller producers will see no influence on the long-term viability of pipelines, distributors, or customers from lowering the price on *their* contracts; besides, many of them have used those contracts for collateral to get the funds they require for further investment. Their bankers may not go along even if they will. Finally, the operator of a gas field typically represents a host of interests—other producers, drilling fund participants, royalty owners, and working-interest owners. Those parties are unlikely to agree universally to give up revenues to which they are contractually entitled. Because these factors will stymie wide-spread renegotiation from the producer side, Tussing and Barlow may well be correct when they state, "Within a few years . . . the most fashionable slogan in the gas industry may well be, 'I can't take and I won't pay. So sue me!' "[13]

To sum up the contracts problem: First, sufficient contracts contain escalation provisions of a form such that a problem exists. Second, there is no voluntary mechanism by which this situation will be resolved without massive amounts of litigation and serious potential harm to the industry as a whole. Third, in the long run it is undoubtedly the case that gas prices will not be held above competitive levels; it is basic economic forces, not the letter of contracts, that will ultimately prevail. But fourth, the lengthy, expensive, and disruptive process this adjustment would require through voluntary agreements and the courts would be inimical to the public interest. Consequently, government action of some sort might be desirable. Fifth, however, such action would have costs and other effects that must be balanced against any benefits from a smoother transition to a less-regulated natural gas market.

The Uneven-Cushion Problem

Will the uneven endowment of regulated (low-priced) gas leave some pipelines (particularly intrastate pipelines) and their customers at an unfair disadvantage under partial decontrol and create market disorder—characterized by differential availability of gas to industrial consumers, by differential profitability of pipelines and distributors, and by differential prices to nonindustrial consumers?

Under the partial deregulation established by the NGPA for 1985, gas purchased under new contracts would cost each field-market buyer the same price, adjusted, of course, for quality differences of various sorts. Newly deregulated gas would be affected by contract provisions, but would (for reasons touched on above) also be at roughly similar prices for different pipelines. Regulated gas would flow to pipelines at prices that depended on the particular mixture of gas of different prices on the system. Small price differences could therefore exist among pipelines for gas in the same regulatory status. The major differences in gas cost among pipelines would arise, however, because they had different proportions of gas supply of each type. If all else stayed the same, therefore, in 1985 under the NGPA pipelines would have average gas costs that varied widely. Intrastate pipelines would have high costs because most of their supplies would command deregulated or unregulated prices. Some interstate pipelines would be in a similar position, while others would depend on high-cost gas for only a small proportion of their supplies.[14]

Of course, everything will not stay the same. The matter of interest is to see what adjustments would occur in this situation and whether those adjustments would yield outcomes inimical to the public interest.

A simple model consisting of two pipelines, one with a "large" regulated-gas cushion and the other with a "small" regulated-gas cushion, can clarify this situation. For reasons discussed above, each pipeline must deliver gas at approximately the adjusted cost of heavy fuel oil to sell gas to bulk heat customers.[15] Similarly, each must pay essentially the same price for newly acquired gas and must choose deregulated contracts to repudiate or honor on roughly the same terms.[16] With different proportions of regulated gas in the two systems, the situation described here is clearly untenable. The rolled-in cost of gas is different for the two pipelines; the rolled-in cost for the small-

17

cushion pipeline is above a level consistent with competing with fuel oil at the burner tip, and the rolled-in cost for the large-cushion pipeline is below that level. Yet the pipelines must both sell gas at the burner tip at roughly the same price. Where does the adjustment take place, and through what mechanism?

The analysis can start at the burner tip. The small-cushion pipeline will seek to shift its full added costs onto consumers and to some extent will initially be successful. Faced with sharply higher gas prices, however, some industrial customers will switch to fuel oil, and sales will be lost. Others will pay higher prices than they would pay under full decontrol and suffer competitive disadvantage until other adjustments occur. Conversely, the large-cushion pipeline will be able to compete better for sales than the small-cushion pipeline and better than it could with full decontrol. Consumers on this system will be benefited by lower gas costs than under full decontrol and lower costs than those paid by competitors on the small-cushion system. On the large-cushion system gas can continue to substitute for oil, and sales may even increase, perhaps through load gains from high-cost pipelines. Such shifts can occur directly when two pipelines serve the same distribution or industrial market and indirectly when activity is depressed in locations where fuel costs are higher in favor of areas where they are lower.

In this discussion it is assumed that with partial decontrol the greater assurance of long-term supply will lead some industrial users to switch to gas because the higher price proves less discouraging than the previous fear of curtailments. (Such switches will occur on large-cushion pipelines, where security of supply and prices below fuel oil costs will be available, but not on small-cushion pipelines, which suffer in both respects.) For simplicity in presentation it is further assumed that field markets will clear at the long-run gas–fuel oil equilibrium price, although there could be a period when the market-clearing price was below that level because buyers could not adjust use patterns instantly.

Loss of sales by the small-cushion pipeline will drive transportation and handling charges per unit up. When those charges are passed through, still fewer sales will result. Conversely, of course, if sales by large-cushion pipelines increase, the opposite effect will occur as fixed costs are spread over more units.

As noted earlier, the rate structures proposed by pipelines will reflect their relative gas costs; the small-cushion pipeline will seek to shift maximum system cost onto those customers with the least elastic demand. To the extent that they are allowed to make such shifts by their regulators, they will avoid losing bulk heat sales that would otherwise disappear. Conversely, the incentives for the large-cushion pipeline to concentrate costs on residential and commercial customers will not be as great. Moreover, regulators need not be as receptive to shifts in rate structures because there is less danger of substantial loss of industrial sales by large-cushion pipelines, with accompanying economic, employment, and load factor consequences.

As the dynamics of this process sort out, the small-cushion pipeline will be unable to sell as much gas as it could before or as it could with full decontrol. In relation to its system size, it will purchase proportionately less unregulated gas and may be unable to honor some newly deregulated contracts at the higher prices. It may also violate take-or-pay provisions for its high-cost contracts. As these reductions in purchases occur, the average purchased-gas cost will fall from its initial state at the time of partial decontrol. Conversely, of course, the large-cushion pipeline will need to meet its expanding sales and to do so must add high-cost gas sources to its system. It can do so and still meet competition at the burner tip. Thus the differences in average purchased-gas costs of the two pipeline systems will narrow through changes in market shares.

Another adjustment process will also occur, with unfortunate consequences for the stockholders of the small-cushion pipeline. The profits of the small-cushion pipeline will be squeezed. Market forces will make it impossible for the pipeline to pass its purchased-gas costs forward in the short run, and the regulatory obligation to serve and long-term contracts may make it impossible in the long run. Linked distributors will be in the same situation. Rates of return allowed by regulatory agencies will be irrelevant if they cannot be earned. Conversely, the large-cushion pipeline will prosper. Thus some of the convergence of pipeline prices at the burner tip and of average gas costs in the field will come from shrunken and engorged stockholder equity.

The uneven-cushion problem will disappear in time even under the NGPA. The old-gas cushion will disappear for all pipelines as fields under regulation are exhausted. In the interim, however, some producers, pipelines, distributors, and consumers will prosper at the expense of others. There will be market disorder. Income distribution questions under alternative control schemes will be discussed further after possible effects on supply of partial decontrol are examined.

The Supply Response Problem

Does the different pattern of field prices under partial rather than full decontrol have a significant effect on the quantity of gas that will be brought to market?

The analysis above suggests that partial decontrol under the NGPA will result in approximately the same average price in the field markets as under full decontrol but that unregulated prices will be above the average and regulated prices below it. In the intermediate run this different pattern of prices can be expected to increase the quantity of gas brought to market. In our model, where gas demand is rather elastic, there will be only limited effects on the market-clearing price.

A decision for partial rather than full decontrol will mean higher prospective prices for new and deregulated gas and a lower price for flowing gas. While hard empirical evidence is lacking, supply elasticity (in the sense of adding to total reserves and long-term deliverability) is probably greater for prospects that have not yet been drilled and for newer fields than for existing wells in old fields. Thus dividing the field market and giving the more elastic segment a higher price and the less elastic a lower one will enhance intermediate-term supplies.

The different time paths of quantities of gas supplied under different control regimes deserve examination, however. With the higher partial decontrol price, previously marginal structures at the extensive margin of exploitation—deep wells, smaller deposits, wells far offshore, and those in deeper waters—will be explored, drilled, and brought into production. Most of the formations made economic at the extensive margin require substantial time to bring to production—perhaps only a matter of months for known small deposits but from one to ten years for the more important projects, such as offshore fields. Thus the full effect of the larger incentives will be delayed. In contrast, once fields are mature (it is from such mature fields that most gas that remains under controls will come), the ability to add reserves in response to a higher price is limited. Output from such fields may be speeded up, however, through such additional investment as infill drilling. It is also true that supply is not totally inelastic; ultimate recovery can be increased. With higher prices, for example, it pays to work over declining wells or to add pumping to bring the flow to pipeline pressure. In the nature of things, efforts of this sort at the intensive exploitation frontier can come to fruition on average more quickly than can efforts at the extensive frontier.

The conclusion is that partial decontrol with its disparity of prices will bring more gas to market than will full decontrol, but that the difference will be greater a year or so after the future pattern of prices becomes known in the industry than it is in the early decontrol period, and less in the more distant future when price paths converge as old gas disappears and the earlier exhaustion of reserves takes effect. This conclusion has significant policy implications that are not all obvious.

Greater supplies will undoubtedly reduce the market-clearing gas price somewhat because gas and fuel oil are not perfect substitutes. Our model suggests, however, that the major effect will be that more gas and less oil will be used in bulk heat applications. While the reductions in oil imports that follow would in themselves be welcome, they will come at the cost of additional resources used up in finding and producing gas that is worth less than it costs. More directly, the resources expended in securing that gas are worth more than the resource cost of oil that it displaces. Hence, although partial decontrol would lead in the intermediate run to more gas industry investment, greater gas supplies, fractionally lower prices, and lower oil imports, this does not mean that the partial decontrol option is necessarily desirable. Income distribution and efficiency effects, both considered below, must also be taken into account.

The Income Distribution Problem

What will be the effects of partial as opposed to full decontrol on the distribution of income among gas industry participants?

Partial decontrol under the NGPA will yield flows of revenues substantially different from those resulting from full decontrol. These differences will be reduced as more of the gas becomes decontrolled, but the effect on market participants in the interim can be substantial. These distributional effects may be modified, of course, by institutional and other factors; specifically, it is unlikely that the full and early adjustments implied by the model used in this discussion will take place. While the magnitudes of the distributional effects may be uncertain, their characteristics seem clear.

Producers as a whole appear likely to be made somewhat worse off by partial than by full decontrol. Partial decontrol may leave average prices below the market-clearing price until adjustments take place and thus hold overall revenues down to some degree. Moreover, since the additional supply that would be forthcoming would arise from investment in high-cost areas, average costs of producing gas would be higher under partial deregulation.

It is clear, however, that producers with mostly unregulated or newly deregulated gas would be better off with partial decontrol. The price they would receive would be above the market-clearing price that would come with full decontrol; conversely, producers of mostly regulated gas will be disadvantaged if the NGPA is maintained in its present form.

Pipelines with small regulated-gas cushions will be disadvantaged because, until adjustments occur, their average gas costs will be higher than under full decontrol and higher than those of other pipelines. Their net revenues and rates of return will be reduced by partial decontrol because of reduced throughput and because they will be unable to pass all costs forward to consumers. Conversely, pipelines with large regulated-gas cushions will be better off with partial than with full decontrol.

Pipelines as a whole will probably be made worse off with partial than with full decontrol. The lower net revenues of small-cushion pipelines are unlikely to be offset by higher profits of large-cushion pipelines. The latter may continue to be limited in their rates of return by regulation and hence may not be able to take full advantage of their privileged position in the field markets. The effect on profits, then, would not be symmetric.

The situation of gas consumers under different decontrol regimes is complex. It depends on where they live, which pipelines and distributors serve them, what decisions are made by regulatory commissions in their jurisdictions, what they use gas for, and, if they are bulk heat consumers, what the specific decontrol regime does to the fuel costs of their competitors. Partial decontrol may lead to slightly lower prices because of adjustment lags, a profit squeeze on small-cushion pipelines, and higher supplies, but these benefits to gas consumers as a class will mostly come at long-run costs to the economy as a whole—of which they are a part. Thus there is no overall *gas* consumer interest that can be defined to favor partial rather than full decontrol, or vice versa. The fact that partial rather than full decontrol leads to waste of resources, however, means that *all* consumers taken together are made worse off (see next section).

One division among gas consumers deserves emphasis. Consumers outside the producing regions would be very heavily favored by partial decontrol, and those in the producing states would be seriously harmed. Under the NGPA, flowing gas previously dedicated to interstate commerce remains under control even after 1985 or 1987. Thus the interstate pipelines will have most of the regulated-gas cushion, and the intrastate pipelines will be paying decontrolled or uncontrolled prices for their supplies.

This discussion of distributional effects has mostly ignored the transitional implications of the two regimes. The existence of reserves in the field markets that can be drawn upon more or less rapidly, of long-term contractual relationships that will slow any adjustments, and of the rigidities inherent in any industry as complex and affected by regulation as natural gas has been noted before. Because of these factors, changes will not come so completely, smoothly, or swiftly as the presentation in this paper would suggest. Rates of change in gas prices may thus be different under partial than under full decontrol depending on the mechanism employed. Although this analysis has concentrated on the results of the two regimes on various parties, the paths of change followed are also important. Matters of degree to the side, however, the distributional effects of partial relative to full decontrol are different, substantially so for some participants. Further, these effects are sometimes different from those commonly supposed, and they work through mechanisms that are not always discernible on casual observation.

The Efficiency Problem

Do the pattern of prices and the behavior of gas producers and consumers under partial as compared with complete decontrol have significant effects on the efficiency with which the economy uses resources to produce goods and services?

The preceding discussion has indicated numerous ways in which prices for gas from different sources and for different uses and consumers will differ between partial decontrol under the NGPA and full decontrol. The changed incentives created by differences in prices will bring about different behavior from most of the participants in the gas market and from those affected by it, such as users

22

and suppliers of competitive fuels. At issue is what the significance of these differences is for the efficiency with which available resources of labor, capital, and materials are transformed into the goods and services that people desire, either for consumption or for investment.

Full understanding of the overall consequences of different systems of decontrol would require treatment within a dynamic general equilibrium framework that took account of all interactions and considered them over time. The time dimension is important because transition states have an important bearing on outcomes. Moreover, when the time dimension is ignored, the capacity of participants to act in expectation of changes may be understated and the diversity of adjustments taking place missed. Unfortunately, such an approach does not seem feasible. The losses in efficiency associated with partial rather than complete decontrol, however, are sufficiently marked that offsetting influences missed because of partial analysis should not alter—except in degree—the conclusion that full decontrol is more efficient. The discussion below illustrates some of the sources of inefficiency in the use of resources that continuation of the NGPA rather than full decontrol would entail.

Wastes of resources in the field markets are simple to identify. As noted previously, under partial decontrol identical gas will be produced at different prices, the average of which would be the price of gas that would be in equilibrium with the price of heavy fuel oil. This means that producers whose properties are regulated could not afford to invest to produce some additional gas even though that gas would cost less to produce than the oil it would displace. Thus gas that is worth to consumers more than it costs is not produced. Under these circumstances some gas is left in the ground in abandoned reservoirs or is wasted in the production process when a full decontrol price would cause it to be brought to market.

In contrast, partial decontrol creates the opportunity to obtain a price for some gas above the decontrol equilibrium level. In response, investment is made to bring to market gas that, at the margin, is worth less to consumers than the resources consumed in producing it. The fact that the *average* price of gas in the field market is roughly the same in both regulatory regimes does not mean that resoures are allocated efficiently. The social loss from the wasted unproduced gas does not offset the social loss from the wasted capital and labor used to produce that which is too expensive; it is added to it.

On the consumption side, the efficiency losses from partial decontrol under the NGPA arise mostly from the uneven endowment of pipelines with regulated gas. Prices to consumers will differ for reasons unrelated to the resource cost of the gas and the cost of delivering it. Customers on large-cushion pipelines (mostly interstate) will therefore consume gas that is worth less to them than the benefit they would get from other uses of the resources going to provide them with the gas service. In contrast, other customers (including most intrastate users) will be induced to use more oil even though at the full decontrol price they would choose to use gas instead. Thus, again, price differences will lead to a lower level of potential social benefit from the resources available to the economy.

The sum of the losses from distortions in the pricing of gas are impossible to calculate. They will not be captured in any conventional macroeconomic model, and they do not show up in measured GNP. While some of the losses are obvious, others are diffused throughout the economy and find their expression in lower real wages and smaller returns to capital. Although precise quantification is not possible, these losses seem far from negligible. Consequently, the relative inefficiency of resource use under NGPA should be considered when policy choices are made.

The Phasing-In Problem

How does phasing in full decontrol between now and 1985 compare with either partial decontrol under the NGPA or full decontrol at that time?

The analysis above of the relative consequences of partial and full decontrol applies in general to the option of phased full decontrol to arrive at the market-clearing price in 1985. Some differences exist, however, and they are explored here.

Phased full decontrol would lead to less of a price shock in 1985, although the interval left to phase in higher prices more rapidly than under the NGPA is limited. The extent to which the price increase is shifted forward, however, would alter the overall effect on market participants and the economy. Gas consumers would be made worse off as a group and producers better off, and the pressures on pipelines and distributors in passing full costs forward would come earlier, but would initially be less intense.

The price index would move up earlier with phased full decontrol, avoiding the jump that otherwise might occur in 1985. This "ramping" of price increases might lessen the macroeconomic adjustment problem; spillover unemployment might be reduced. Whether this is a significant benefit depends, of course, on the size of the shock that would otherwise occur. It also depends on whether expectations of future price increases would be sufficiently established to induce optimal adjustments in any event. Overall, unless the price difference at the end of 1984 would be greater than suggested in this analysis, it appears that there would be little advantage to phasing in from a macroeconomic perspective.

The decision on gradual full decontrol rather than the NGPA thus rests largely on the same issues that divide the NGPA and instant full decontrol in 1985. This conclusion suggests that *which* prices are increased how fast with phased decontrol is of great significance. Under the NGPA some prices will be rising fast and others more slowly or not at all in real terms, with differential distributional effects and with harmful effects on efficient resource use. Phased full decontrol would have different prices rising at yet other rates, with still other effects on efficiency. Another option is to retain regulation but to move prices higher selectively and on a different schedule. The Federal Energy Regulatory Commission has offered such alternatives for comment. Finally, a schedule gradually removing ceiling prices on all gas, perhaps at different rates for gas in different categories, could be established through new legislation. Each of these options would require individual analysis; except for the avoidance of a possible price shock, no generalizations can be made about phased decontrol relative to NGPA or complete (but anticipated) decontrol.

Conclusions: Uncertainties and Issues

The discussion above has offered a simple model of price formation in the natural gas field markets. This model formed the basis for an approach to a number of issues that bear on the decisions that could be made on questions of control and decontrol of natural gas.

The basic questions to which the analyst must repeatedly return are how fast, how completely, and with how much ancillary disruption adjustments to changed regulatory and market conditions will occur. If the adjustment is relatively rapid and complete, a simple economics-oriented model such as the one presented here will be a reasonable guide. It will indicate the directions of the changes that will occur and the approximate distance those changes will go. If the basic model is sound, new evidence or different interpretations may change the predicted outcomes, but not the process by which the predictions are made. If, however, the system for natural gas production and use is so rigid (because of contractual, institutional, regulatory, or technical characteristics) that change is stymied or moved into other channels, then no straightforward predictions can be made.

The conclusion reached here is that although rigidities exist, the framework presented provides a reasonable guide to outcomes that can be expected from alternative regulatory regimes. Further work remains to be done in refining and determining the assumptions and information to be used with the framework to arrive at reliable predictions.

The results of this analysis can contribute to the coming policy debate, but by themselves will not resolve it. The question Congress and the administration must answer is, Weighing anticipated effects on overall output, income distribution, and other implications of different regulatory regimes, which sets of outcomes are preferable?

Notes

1. P.L. 95-621, 15 U.S.C. 717ff.

2. The bibliography at the end of this paper is a selection from the recent "ephemeral" literature that has grown up around the natural gas regulation question, emphasizing items that are likely to be found only in specialized collections.

3. U.S. Department of Energy, Energy Information Administration, *Annual Report to Congress 1980,* vol. 2 (Washington, D.C., 1981); and U.S. Energy Information Administration, Natural Gas Division, Office of Oil and Gas, "An Analysis of the Natural Gas Policy Act and Several Alternatives, Part 1, The Current State of the Natural Gas Market," prepublication draft (Washington, D.C., 1981), p. 95.

4. Natural Gas Policy Act of 1978, secs. 201–8, and Power Plant and Industrial Fuel Use Act of 1978 (P.L. 95-620, 42 U.S.C. 8301ff.).

5. Energy Information Administration, "Analysis of the Natural Gas Policy Act," p. 63.

6. Catherine Good Abbott, "Issues in the Debate over the Natural Gas Policy Act of 1978" (Washington, D.C.: DOE, Office of Policy, Planning, and Analysis, 1981), p. 6; Interstate Natural Gas Association of America, "Analysis of Natural Gas Decontrol" (1981), pp. iv, 4; and Benjamin Schlesinger et al., "Consumer Impact of Indefinite Gas Price Escalator Clauses under Alternative Decontrol Plans," *Gas Energy Review* (December 1981), p. 8.

7. Schlesinger et al., "Consumer Impact," p. 6; and Catherine Good Abbott, "Impacts of Alternative Decontrol Strategies on Natural Gas Ratemaking" (Washington, D.C.: DOE, Office of Natural Gas, Office of Policy and Evaluation, 1981), p. 2.

8. American Gas Association, *Gas Facts 1980* (Arlington, Va.: AGA, 1981); and *Monthly Energy Review* (December 1981).

9. *Monthly Energy Review* (December 1981, September 1979); *Oil Daily,* February 22, 1982; *Economic Report of the President 1982* (Washington, D.C., 1982).

10. Ignored in this conclusion is the possibility that the Federal Energy Regulatory Commission (FERC) will act on its own to raise average prices by adding new high-priced categories and allowing higher prices for old gas. U.S. Federal Energy Regulatory Commission, "High-Cost Natural Gas Produced from Intermediate Drilling, Notice of Proposed Rulemaking," Docket No. RM82-8 (Washington, D.C., December 30, 1981); and FERC, "Legal Bases for Eliminating Price Vintaging of Interstate Flowing Gas" (Washington, D.C., November 1981).

11. For discussion of the contracts issue, see DOE, Energy Information Administration, Office of Oil and Gas, *Natural Gas Pipeline/Producer Contracts: A Preliminary Analysis* (DOE/EIA, Washington, D.C., December 1981); Decision Analysis Corporation, *Analysis of Natural Gas Producer/Interstate Pipeline Contracts* (Annandale, Va.: Decision Analysis Corporation, July 1, 1981); Steve Watson, "Natural Gas Contract Structures Affecting Decontrol" (Paper presented at Operations Research Society of America, Houston, Texas, October 13, 1981); Schlesinger et al., "Consumer Impact"; and Edmond R. Dupont and Associates, "Preliminary Assessment of Gas Rate Changes under NGPA" (Washington, D.C., June 26, 1981).

12. Arlon R. Tussing and Connie C. Barlow, "The Rise and Fall of Regulation in the Natural Gas Industry," *Public Utilities Fortnightly,* vol. 109, no. 5 (March 4, 1982), p. 18.

13. Ibid.

14. Evidence suggests that in late 1981 gas that would not be subject to decontrol under the NGPA made up almost 80 percent of the throughput of some interstate pipelines and less than 30 percent of that of others. Energy Information Administration, "Analysis of the Natural Gas Policy Act," p. 66.

15. As partial decontrol occurs, gas prices will rise to all purchasers, making gas less competitive than before in all markets. This discussion, however, compares the *relative* position each pipeline will occupy. While the large-cushion pipelines' purchased-gas cost will increase, the fact that small-cushion pipelines will bear more of the uncontrolled gas costs means that the average price of large-cushion pipelines will still be lower than it would be under full decontrol. The uneven cushion can also have an effect on the market-clearing price, but this matter is ignored in this discussion.

16. Distributors receiving gas from pipelines of each type are in positions analogous to their supplier pipelines. The discussion here by implication treats the downstream delivery system as an integrated whole, but the story is carried by tracing effects on pipelines.

Bibliography

Abbott, Catherine Good. "Impacts of Alternative Decontrol Strategies on Natural Gas Ratemaking." Washington, D.C.: U.S. Department of Energy, Office of Natural Gas, Office of Policy and Evaluation, February 1981.
———. "Issues in the Debate over the Natural Gas Policy Act of 1978." Washington, D.C.: U.S. Department of Energy, Office of Policy, Planning, and Analysis, November 1981.
———. "The Natural Gas Market in the Transition to Decontrol." Washington, D.C.: U.S. Department of Energy, Office of Policy and Evaluation, June 1980.
———. "Pitfalls on the Road to Decontrol: Lessons from the Natural Gas Policy Act of 1978." Washington, D.C.: U.S. Department of Energy, Office of Policy, Planning, and Analysis, Division of Energy Deregulation, March 1982.
American Association of Petroleum Geologists, Division of Professional Affairs. "Natural Gas Price Decontrol Position Paper." January 1982.
American Gas Association. "Overview of the Alaskan Natural Gas Transportation System." *Gas Energy Review* (January 1982).
Bethell, Tom. "The Gas Price Fixers." *Harper's* (June 1979).
Carlson, Mary et al. "Intrastate and Interstate Supply Markets under the Natural Gas Policy Act." Washington, D.C.: U.S. Department of Energy, Energy Information Administration, May 1981.
Cicchetti, Charles J. "The Tax and Revenue Effects of Natural Gas Deregulation." Madison Consulting Group for the Northeast Coalition for Energy Equity, October 29, 1981.
Decision Analysis Corporation. *Analysis of Natural Gas Producer/Interstate Pipeline Contracts.* Annandale, Va., July 1, 1981.
Demarest, William F., Jr. "NGPA Revisited: Implications of the Historical Origins of the Market Order Issue." Paper presented to the International Association of Energy Economists. Houston, Tex., November 13, 1981.
Dingell, John D. "Speech of the Honorable John D. Dingell before the Federal Energy Bar Association," January 29, 1981.
Dupont, Edmond R., and Associates. "Preliminary Assessment of Gas Rate Changes under NGPA." Washington, D.C., June 26, 1981.
Easterbrook, Gregg. "The Energy Crisis, How to End It: Natural Gas." *Washington Monthly* (October 1980): 20–30.
Energy Action. "An Estimate of the Consumer Cost of the Administration's Proposal to Increase the Price for So-called Intermediate Deep Natural Gas." A Report of the Energy Action Project of the Citizen/Labor Energy Coalition. Washington, D.C., January 18, 1982.
Erickson, Edward W. "Myth and Reality: A Retrospective and Prospective View of the NGPA." Remarks to the National Association of Petroleum Investment Analysts, October 1981.

_____. *Natural Gas and the Natural Gas Policy Act: A Pragmatic Analysis*. Washington, D.C.: Natural Gas Supply Association, January 1981.

Fleming, Russell Jr., and Joseph M. Oliver, Jr. "The Gas Distributor Approaches Deregulation." *Public Utilities Fortnightly* 108 (July 2, 1981).

Foster Associates, Inc. "A Comparison and Appraisal of Ten Natural Gas Deregulation Studies." Prepared for the Chemical Manufacturers Association. Washington, D.C., February 16, 1982.

_____. "A Report on: Basic Factors Influencing Gas Availability to Industrial Consumers in the Interstate Market and Projected Purchased Gas Costs of Interstate Pipelines." Prepared for the Petrochemical Energy Group. Washington, D.C., September 17, 1980.

Gardner, Bruce. "The Impact of Natural Gas Deregulation on the Farm and Food Sector." University of Maryland, January 1982.

GHK Companies. "Analysis of 'A Study of Alternatives to the Natural Gas Policy Act of 1978' (U.S. Department of Energy, November 1981)." Oklahoma City, Okla.: GHK Companies, November 1981.

Hill, Daniel H. "The Impact of Natural Gas Deregulation on American Families." *Economic Outlook USA* (Autumn 1981): 84–87.

Interstate Natural Gas Association of America. "Analysis of Natural Gas Decontrol." Washington, D.C., December 1, 1981.

_____. "Memorandum to: Tax Committee, Gas Requirements Committee, Finance Committee, Government Affairs Committee, Synthetic Gas Committee, Legal Committee, and Rate Committee." Washington, D.C., September 9, 1981.

_____. "Supplemental Statement before the Senate Committee on Energy and Natural Resources on Implementation of Title I of the Natural Gas Policy Act." Washington, D.C., December 7, 1981.

Jacoby, Henry D., and Arthur W. Wright. "Obvious and Not-So-Obvious Issues in Natural Gas Deregulation." Massachusetts Institute of Technology, November 4, 1981.

Jimison, John W., and Lawrence Kumins. "The Natural Gas Policy Act: Reform Revisited." Issue Brief No. IB81020. Washington, D.C.: U.S. Library of Congress, Congressional Research Service, Environment and Natural Resources Policy Division, December 1981.

Lareau, Thomas J. "The Impact of Oil Market Disruptions on Other Fuel Prices." Discussion Paper D-82D. Washington, D.C.: Resources for the Future, Center for Energy Policy Research, March 1982.

Lauter, Dave. "The Natural Gas Debate Takes Shape." *Fact Sheet*. Washington, D.C.: U.S. Congress, Environmental Study Conference, February 4, 1981.

Lichtblau, John. "U.S. Natural Gas Policy and Outlook for Gas Imports." Paper delivered at the International Seminar on Natural Gas and Economic Development, Cairo, Egypt. New York: Petroleum Industry Research Foundation, February 26, 1982.

Loury, Glenn C. *An Analysis of the Efficiency and Inflationary Impact of the Decontrol of Natural Gas Prices*. Washington, D.C.: Natural Gas Supply Association, April 1981.

29

Means, Robert C. "A Brief Comparison of Energy Information Administration, *Analysis of Economic Effects of Accelerated Deregulation of Natural Gas Prices* (August 1981), and U.S. Department of Energy, Office of Policy, Planning, and Analysis, *A Study of Alternatives to the Natural Gas Policy Act of 1978* (November 1981)." Washington, D.C.: Federal Energy Regulatory Commission, Office of Regulatory Analysis, December 3, 1981.

_____. "A Preliminary Analysis of the Natural Gas Market-Ordering Problem." Austin: University of Texas, Center for Energy Studies, February 1981.

_____. "Speech before the Ninth Annual Illinois Energy Conference," October 28, 1981.

Merrill, Peter R. "The Regulation and Deregulation of Natural Gas in the U.S. (1938–1985)." E-80-13. Harvard University, Energy and Environmental Policy Center, John F. Kennedy School of Government, January 1981.

Murphy, Frederic H., Richard P. O'Neill, and Mark Rodekohr. "An Overview of Natural Gas Markets." *Monthly Energy Review* (December 1981): i–viii.

Muzzo, Steven E. "The Effects of Natural Gas Decontrol Policy Options on Future Domestic Gas Supplies." Paper presented to the International Association of Energy Economists, Houston, Texas, November 12–13, 1981.

O'Neill, Richard P. "The Interstate and Intrastate Natural Gas Markets." *Monthly Energy Review* (January 1982): i–ix.

_____. "Natural Gas Drilling and Production under the Natural Gas Policy Act." *Monthly Energy Review* (February 1982): i–viii.

Process Gas Consumers Group. *The Case for Repeal of Incremental Pricing of Natural Gas*. Washington, D.C., November 1980.

_____. *Cost-of-Service Ratemaking: Making Natural Gas Rates Equitable*. Washington, D.C., May 1981.

_____. *End-Use Curtailment of Natural Gas: A Question of Priorities*. Washington, D.C., May 1981.

_____. "Legislative Specifications for the Natural Gas Policy Reform Act of 1981." Washington, D.C., September 17, 1981.

_____. *Marginal Cost Pricing: Theory vs. Reality*. Washington, D.C., May 1981.

Reynolds, Alan. "Decontrolling Natural Gas: The Issues." Morristown, N.J.: Polyconomics, Inc., August 31, 1981.

_____. "Freedom for Natural Gas." *Petroleum Independent* (October 1981).

Schlesinger, Benjamin et al. "Consumer Impact of Indefinite Gas Price Escalator Clauses under Alternative Decontrol Plans." *Gas Energy Review* (December 1981).

Solomon, Burt. "Gas Decontrol: Will It Wait 'Til Next Year?" *Energy Daily* 10 (January 28, 1982).

Stalon, Charles G. "Deregulation of Wellhead Prices: For What Purpose?" Paper presented to the Ninety-third Annual Convention, National Association of Regulatory Utility Commissioners, San Francisco, November 16, 1981.

Standard Oil Company of Indiana. "Natural Gas Decontrol and You." Draft, 1982.

Tiano, J. Richard. "'Market Ordering' Devices from a Gas Distributor's Viewpoint." *Public Utilities Fortnightly* 108 (March 26, 1981).

Tussing, Arlon R., and Connie C. Barlow. "The Rise, Maturity and Fall of Regulation in the Natural Gas Industry." *Public Utilities Fortnightly* 109 (March 4, 1982).

United Distribution Companies. "To Make Deregulation Work in the Public Interest, the Natural Gas Policy Act of 1978 *Must Be Amended.*" Detroit, December 22, 1981.

U.S. Department of Energy, Energy Information Administration. "Alternative Fuel Price Ceiling, Price Threshold on High-Cost Gas." *Energy Clearinghouse* (January 29, 1982).

————. "Analysis of Economic Effects of Accelerated Deregulation of Natural Gas Prices," prepublication draft. Washington, D.C., August 1981.

U.S. Department of Energy, Energy Information Administration, Office of Oil and Gas. "An Analysis of the Natural Gas Policy Act and Several Alternatives: Part I, The Current State of the Natural Gas Market," prepublication draft. Washington, D.C., December 1981.

————. *Intrastate and Interstate Supply Markets under the Natural Gas Policy Act.* Washington, D.C., October 1981.

————. *Natural Gas Pipeline/Producer Contracts: A Preliminary Analysis.* Washington, D.C., December 1981.

————. *Natural Gas Supply through 1985.* Washington, D.C., February 1982.

U.S. Department of Energy, Office of Policy and Evaluation, *Technical Staff Analysis in Response to Notice of Proposed Rulemaking on Phase II of Incremental Pricing.* Washington, D.C., February 8, 1980.

U.S. Department of Energy, Office of Policy, Planning, and Analysis, Division of Energy Deregulation. *A Study of Alternatives to the Natural Gas Policy Act of 1978.* Washington, D.C., November 1981.

————. "Two-Market Analysis of Natural Gas Decontrol." Appendix A to *A Study of Alternatives to the Natural Gas Policy Act of 1978.* Washington, D.C., November 1981.

————. "Industrial Gas Demand." Appendix B to *A Study of Alternatives to the Natural Gas Policy Act of 1978.* Washington, D.C., November 1981.

————. "Macroeconomic Consequences of Natural Gas Decontrol." Appendix C to *A Study of Alternatives to the Natural Gas Policy Act of 1978.* Washington, D.C., November 1981.

U.S. Federal Energy Regulatory Commission. "High-Cost Natural Gas Produced from Intermediate Deep Drilling, Notice of Proposed Rulemaking," Docket No. RM82-8. Washington, D.C., December 30, 1981.

————. "Legal Bases for Eliminating Price Vintaging of Interstate Flowing Gas." Washington, D.C., November 10, 1981.

U.S. Senate. *Implementation of Title I of the Natural Gas Policy Act of 1978.* Hearings before the Committee on Energy and Natural Resources, November 5–6, 1981.

Watson, Steve. "Natural Gas Contract Structures Affecting Decontrol." Paper presented to the Operations Research Society of America, Houston, Texas, October 13, 1981.

Williams Companies. "Natural Gas Deregulation and the Intrastate Natural Gas Industry." Tulsa, Okla., August 25, 1981.

Commentary

Edward W. Erickson

Milton Russell as usual has provided us with a very provocative piece of work. I certainly agree that for the foreseeable future the market-clearing mechanism for natural gas will be driven by residual fuel oil prices, but I would qualify that by saying that one has to be concerned about what residual oil and where. There is high- and low-sulfur residual. The low-sulfur residual is currently selling at about 100 percent parity with crude oil in New York harbor before movement to the burner tip. Higher-sulfur residual is selling at a discount in the Gulf Coast. So we must look at those relationships before we can arrive at an overall judgment that, say, 70 percent of the crude oil price is a good number for natural gas prices. I agree with Russell that in the event of substantial deregulation, a very large price movement would take place.

There are many ways to talk about the persistent shortage of reserve additions that has resulted from restrictive ceiling prices in natural gas field markets. Since this is a group of economists, that piece of evidence alone ought to persuade everybody that the natural gas market is not clearing on a reserves addition basis. The fact that gas competes at the burner tip with alternative fuels and that regulatory policies are very important to interfuel competition is the subject for another paper. There is going to be a considerable need—in the current environment, any extension of the current environment, a completely deregulated environment, or a partially deregulated environment—for public service commissions in the various states to adopt imaginative policies to keep the industrial load on the system. Policy makers must avoid what looks like short-run good sense—overcharging industrial users to subsidize residential and small commercial users—if we are not to penalize the whole system through loss of load factor.

I noted with interest Russell's adoption of the position that consumer prices are not affected by whether decontrol is full or partial. That is not a proposition I am willing to swallow whole, in part because it relates to my previous point with regard to how individual state utility commissions respond and also because it is not clear to me that the simple-minded approach to what supply response will be is the appropriate way to look at a bifurcated price structure. It seems to me that what fundamentally drives things, what the

33

fundamental proposition is that we have to come to grips with—and I think that it can be demonstrated in any number of ways—is that there is a shortage of reserve additions and will probably continue to be one under the NGPA as it is or extended into the middle 1980s and beyond. And that shortage of reserve additions drives the contract problem, the market-ordering problem, and the current distortions in the prices of deep gas.

Although we could fundamentally agree on the structure of the market-clearing mechanism, I think it is dangerous to pretend to know where that price is or will be. As the best evidence of that, I offer the NGPA itself. For the market to clear, supply response is the critical issue. The NGPA in its current form, even if deregulation provisions are allowed to operate in 1985, is not likely to generate a sufficient supply response or the kind of supply response that I think Russell has in mind.

There is a way to test his hypothesis, a way that looks unfortunately increasingly likely. That is to let the NGPA run its course and see whether, in anticipation of higher prices in 1985, a drilling boom occurs in 1983 and 1984—which, for the usual discounted-cash-flow, return-on-investment reasons, would be a likely result. If we structure this question as a bet, I will take the side of the bet that the drilling boom will not materialize, that ARMCO will not decide to speed up its pipe mill in anticipation of the wonderful effects of the NGPA, and that—since gas is found by drilling holes in the ground—in the absence of a natural gas drilling boom, we will perpetuate the shortage of reserve additions into the latter half of the 1980s.

The contract problem is a result of the competitiveness of field markets for natural gas. Even the Federal Trade Commission is on record about that. In competitive markets, the effects of restrictive ceiling prices are not surprising. People try to find other ways to compete. Competition among pipelines for the limited gas supplies that are available has led them, since they are unable to compete as they would in a normal market, an unrestricted market—directly through price provisions—to compete in promises. The contract problem can be summarized as "promise them anything, but give them Arpège"—where Arpège is the NGPA-regulated ceiling prices. I have no reason to believe that pipeline managements are somehow fundamentally defective as market operators. In fact, the ingenuity they have displayed in finding ways to compete for the limited supplies of reserve additions that are available itself supports my faith that under complete decontrol they would find ways to play off the producers against one another in order to renegotiate very quickly and expeditiously the contracts that some people are now worried about.

Although many pipelines have large numbers of contracts, several thousand in some instances, the distribution of those contracts by size, like the distribution of resources in nature, is approximately log normal. I know of one pipeline with nearly 3,000 contracts that has more than 90 percent of its gas in fewer than 200 contracts, basically with a well-identified set of producers, so I do not see the renegotiation problem as insurmountable. I see competition among producers as leading to a fairly expeditious renegotiation, where that renegotiation would be in the competition to sell new gas supplies at market-clearing prices. Just as contract terms have been levered up because of a shortage of reserve additions, they will be levered down to the same market-clearing prices that new gas supplies are selling for. Likewise, to maintain rates of take above minimum levels, we can expect to see considerable flexibility on existing contracts as well. So, although Russell used the term "long run" and that gives us a feeling of distant time, the long run, as we know in economics, is associated with no specific calendar time. It depends on the circumstances necessary for the full adjustments to take place, and I believe the long run for this issue is somewhat less than a year.

I agree with Russell that the market-ordering problem can be aptly characterized as a market-share problem. There *is* a market-ordering problem, of which deep gas prices are a symptom, but it takes only a little arithmetic to discover that deep gas prices are not sufficiently high to validate the full hypothesis. With a substantial supply response, pipelines will pay no more than they have to pay for gas; so an adequate supply response in conjunction with deregulated prices is the key to a solution to the market-ordering problem. To the extent that we continue to have a shortage of reserve additions, competition for the limited reserves available will exacerbate the market-ordering problem as Russell defines it in terms of market share.

On this supply response issue, will producers respond to high market conditions and hothouse prices under the kind of scenario for the playing out of the NGPA that Russell describes? Or will they, on the basis of prudent judgment and past experience, respond to some substantial discount from those nominal prices? There is also an interaction between how producers perceive those prices and the signals they get from intervention in the market to, say, impose a cap on contract provisions and some kind of a legislative solution to the contract problem. If those things go together and producers who have been active for the last decade or more look back on the history of vintaging and on crude oil price controls, it is not encouraging to believe that there will be a full response to the nominal prices. We have a history of new, new new, new new new, new new new new gas. If we add to that an indication of legislative willingness to intervene in contracts arrived at between willing buyers and willing sellers, those prices will be costly for consumers, but they will not elicit the kind of supply response that they would in another kind of policy environment.

With regard to economic efficiency, I think Russell hit that nail very nicely on the head. The question of how fast we should move to correct the fundamental deficiencies in the institutions and regulatory constraints that impinge on the gas market is of considerable interest to me. Not long ago I thought we were going to move quickly enough in that direction so that I could stop worrying about natural gas and take up an alternative profession. But it looks as if not only my children, but maybe even my grandchildren, will go to college on the natural gas issue.

Benjamin Schlesinger

The first reaction I have to Milton Russell's very thoughtful paper is that I kept turning the pages back and ahead as I read and reread it; in so doing, I discovered that the paper has something for everyone. On an early page, there is a misstatement about the contracts problem that is corrected on a later page. The same comment applies to the consumer benefits of partial decontrol. The paper was a very appropriate springboard for this conversation, but let me bring a couple of other matters into the conversation. Probably the most important of these is the position of the American Gas Association (AGA) which has, I think, been maligned and misunderstood. As probably best laid out in the letter of AGA president George Lawrence to the *Oil and Gas Journal* last month, we are and have been advocates of decontrol of natural gas prices since the early 1970s. Our effort is to ensure a smooth transition to decontrol. We want to see lots of gas—lots of natural gas and lots of supplemental gas— because our motive and that of our customers is to have adequate and reasonably priced gas supplies. Which producers benefit is a secondary issue to us, so that whether we advocate partial or total decontrol depends strictly on our interests and the interests of our customers.

One reporter pointed out that the gas pipelines and, I think by extension, the gas distribution systems are literally and figuratively in the middle of the issue. I think that is well put. If anything, we are a bit closer to the consumer viewpoint than to the producer viewpoint, although I have yet to figure out what the single producer viewpoint really is. Clearly, there are differences. But if I had to put us close to one side or the other, it would be the consumer side. Various proposals force us to become directly and personally involved with consumers' concerns. Uncollectables now approach 2 percent of revenues, and that is going to increase as prices increase. Our concerns are gas utilities, the concerns of corporations who cannot "get up and move away." Your gas company is not about to leave a central city and move somewhere else in response to more favorable economic conditions. They cannot pick up the pipelines from Brooklyn and move them to Arizona. We are there to stay and to face the consumers. So we are closer to them.

We strongly endorsed the president's decision not to propose an accelerated natural gas decontrol bill this year. It is the right decision given the national priorities. But we urged the administration to continue to bring together all segments of the gas industry—including consumer groups and industrial gas users—to work out the necessary modifications of the NGPA. We do advocate modifications of the NGPA, which must be thoughtful and timely to achieve the common objective of an orderly transition to deregulation of natural gas wellhead prices.

There has been considerable progress in this debate. Discussion began last year with proposals to throw the cards up in the air and let them fall where they may: decontrol everything immediately. That was soon abandoned. We got on to a three-year decontrol path that still ignored the problem of the contracts, and now we are struggling with how best to ensure a smooth decontrol track, looking at all the factors.

Regarding the contracts, I agree with Russell. They will cause mischief in the short run, but the specific kind of mischief—I am speaking here as an engineer—the "transient effect" that these contract escalator clauses will cause is one that forces consumers and buyers to argue prices down, not producers to argue prices up. The transient effect is a price overshoot; I will not even call it that: it is an upshoot, in which gas prices will be higher than they ought to be. We buyers of gas—and the AGA represents companies who buy 75 percent of all the gas produced at the wellhead—and consumers do not want to see this transient overshoot of market conditions. It is as simple as that. If we expected an undershoot of some kind, I suppose we all would be talking about this from different sides of our concerns. The problem is one of overshoot. How long will it last? We would be less than responsible if we just assumed that it will go away quickly. We have heard that some high-placed officials and others, including some producers, appear to believe that these high prices will fall right away, but I do not think that likely; I think it will take time to negotiate them out of the system. Russell correctly cites by inference the example of the gas producer: how is he going to tell his stockholders that he just gave away $20 million in the interest of "market clearing"?

Now let me turn to market-clearing conditions. There has been a great deal of confusion about this, and there is a point I need to clarify in Russell's paper. We by no means believe that the market-clearing level for gas prices in a totally free economy is 70 percent of the price of crude oil. As the market is structured today, it is considerably less than that. The kind of restructuring the market would have to go through to get the price up to 70 percent of crude oil at the wellhead may occur in the future. Market clearing is very much a function of time and of the structure of the industry. Our own analysis indicates the market-clearing price is now less than 60 percent of the crude oil price. This agrees with several other forecasts to the year 1985. Wellhead gas prices are now 40 percent of crude oil prices on the average and 55 percent on the margin. Oil prices are at $32 a barrel. Section 102 gas wellhead prices are currently $3.06 per million Btu, which is·55 percent of the price of crude. Although the average is about 40 percent, with dipping oil prices—should they fall as far as some think—gas will lose load. The gas industry will serve a little less as oil prices drop through their present levels. I think that demonstrates how close we are now to market-clearing conditions. The best oil analysts seem to be saying that prices will fall to $30 or $31 per barrel—perhaps $32, perhaps less than $30—and stay there for the next year or year and a half. The forces seem to be in motion, optimists believe, so that the price will then go back up. In fact, the whole rationale for accelerating decontrol, the prospect of a monstrous fly-up from the current prices to some theoretically high market-clearing level, is just not there. We are practically at the market-clearing level now. There is the prospect of a fly-up as a result of the transient effect of the existing contracts but not as a result of some average fly-up to market-clearing level.

As for the unevenness of the cushion in the event of partial decontrol, the whole problem of partial decontrol is one of choice. We believe that the resource base distribution has been accurately identified by the Potential Gas Committee, whose estimates are close to those of the U.S. Geological Survey: they are now within 17 percent of each other with regard to the remaining undiscovered recoverable resource base. In light of what geologists believe about the distribution of the resource base, the prospect for increased reserve additions is greater given the split market, with some old gas in the system. The higher marginal price will send drillers into just the right areas, as it has done in the recent past. That does not mean that the price will go up and up as a function of an unequal cushion distribution. In fact, the very opposite is true, as Erickson rightly points out: the pipelines are in the deep areas buying up reserves because there is such a short life on the new gas. Every time a pipeline goes out to buy section 102 gas, its reserve life index drops a little more, and we all know how important that is to the banking and financial community. Section 107 prices have now been stabilized and are slipping in some areas, according to the most recent monthly reports. This may well be a function of the softness in oil prices as producers aim more and more for gas.

One point that Erickson brought out—that there is a persistent shortage of reserve additions—I think is not borne out by the most recent statistics, which indicate that in 1981 gas well completions were up 14 percent over 1980, which is an "increase in the increase" in completions, the first since the NGPA was enacted. There is plenty of gas well drilling out there. Reserve additions have come in during the last year at from 70 to 90 percent of gas production, depending on how the gas total is defined. Basically additions are at about 90 percent, which is close to replacement. Under the present law and climate, drilling will tip more toward gas (or will have tipped more toward gas this year and next) because of low oil prices. Oil prices overshot after oil decontrol, just as we are concerned that gas prices will overshoot. They overshot to $40 on the spot market a year and a half ago and have been slipping and fighting the demand battle ever since.

And that is where I want to end my comments today: on gas demand. I see demand, rather than partial or total decontrol as the central issue of gas policy. Our industry's efforts on decontrol and our questions about the contracts ought to continue so as to get these questions resolved, because uncertainty itself is probably the industry's worst problem in trying to sell gas. Uncertainty in the price path also contributes to the drilling hesitancy. Some producers are probably concerned that controls could be reimposed. The point, however, is that we do not really know what gas demand is in this country. Gas companies are not free to create the marketing staffs, marketing capacity, and marketing acumen and to build the marketing capability to sell gas in all traditional and nontraditional uses. There are severe constraints at both the state and the federal levels. These have grown up under conditions that made them appear appropriate at the time. They are not appropriate now. Although highly efficient gas furnaces are on the market, some gas companies are not even aware of what is available. We are making it a large part of our business to communicate with our own companies about what is there. There is no such thing as a gas appliance manufacturer: most of them are electrical appliance manufacturers who also make some gas-using products. In the industrial market, we have not even begun to explore some of the prospects for increased use of gas that may well be there. Gas-fired cogeneration is stymied by a wholly artificial regulatory approach. If electric companies can be shown that it is in their interest to take a small, bite-size chunk of electric power rather than some giant chunk that may or may not be there or be needed in ten years, we may well have a market for gas-fired cogeneration. Natural gas in automobiles was a joke three years ago; it is not a joke now. It will take us time and common effort and appreciation for the role gas can play to build these kinds of markets. We cannot do that with the prospect of a transient effect that will push gas prices far above market levels for a period of time, whether it is one, two, or three years. That inhibits our marketing, and it will create a climate for reimposition of demand controls, which is just the kind of thing we want to avoid.

Discussion

DR. RUSSELL: I couldn't agree more with Ben Schlesinger when he says that we don't know what gas demand is—that is, what quantity of gas will be demanded under different circumstances. I simply suggest that when we start talking about what that market-clearing price will be, whether it will be 40 percent or 55 percent or 60 percent, when those estimates are based on a given demand curve rather than taking into account the shift in the demand curve outward that may well occur with greater certainties of supply on the part of industrial consumers, it seems to me that the view that gas prices won't track and move very rapidly upward comes into some question.

Moving from this demand question, let's talk about supply in terms of price, which ties in with comments by both Schlesinger and Erickson. Where gas is competing with oil in a relatively large market for bulk heat, and if that market is such that relatively small changes in utility prices would have an important effect on the market share of oil versus gas, then in a way we can say that the quantity of gas supplied will not have much effect on that price unless it moves outside those two boundaries. Consequently, I think it is certainly true that the supply response from partial decontrol will be different from that of full decontrol, but the different supply response does not mean that the prices will be substantially different under the two regimes. I think the analysis works out that way. We all know that gas and oil are not perfect substitutes, which is not to say that there won't be some price difference, but not a large price difference.

That leads me to the question of the quantity of gas supplied and how important that is. Schlesinger and Erickson talked about the desirability of having more gas, that what the gas distribution and pipeline industries want and what gas consumers want is more gas. As the pipelines and distribution companies are, of course, interested in more gas, and if more gas is obtained at reasonable cost, consumers would also be interested in more gas. The question is, are we creating a distorted situation where people are using more gas that is worth less to them than the resources the economy as a whole must give up to produce it? It sounds very good to say that the desirable goal is to have more gas because if we have more gas we have more investment in the gas industry, which is good; we'll get more spillover effects into the oil industry, which is good; oil imports will fall, which is good; and we'll have improvements in the balance of payments, which is good. But all of that comes at a cost to the economy as a whole of a less efficient use of resources. All too often we live in a world of partial equilibrium—certainly the paper I wrote is in a partial equilibrium—but we need to broaden our views a bit and ask ourselves what the overall effects of these decisions are. In this case, we raise the question whether more gas is necessarily good for us.

GORDON GOOCH: I would like to discuss a couple of points, if I may. First, I think Russell's presentation was excellent. I would like to suggest a few more subsets for your consideration.

When we look at the industrial users, I think it's a mistake to think too much about the industrial boiler fueler and utility boiler fueler with dual-fired capabilities, because there are other sets out there. There are the feedstock and process and plant protection sets of industrial users that do not have dual-fired feedstock capability or dual-fired process capability or dual-fired plant protection capability. Some of them do not have even dual-fired boiler capability. Those plants cannot move from one region of the country to another, or even next door to another pipeline; they cannot switch necessarily between Texas Eastern and Transco or Tennessee, even if they are in the same market area, to follow a gas supply. The real hostages of the situation, or among the hostages of the situation, are what might be called high-priority industrial electric utility users that have no option. The suggestion that that can somehow be mitigated by a change in rate design is no comfort, because if the mitigation of rate design is to keep the dual-fired dump interruptable user on the system, the feedstock process user finds that he will be subsidizing the boiler fuel user on the system. I suggest that some consideration be given to those subsets of problems. The industrial utility users are smart enough to see what the respective cushions of their pipelines are and are smart enough to understand what Dr. Schlesinger said—maintain price controls, protect the cushion, keep prices from rising. We can reasonably project the consequences of that.

I would like to add a couple more subsets in another subject area. In the contract area, I suggest that you direct your attention also to the problem of area rate clauses. There are producer contracts that were required by federal regulation in the 1960s that will have only area rate clauses, other than definite price escalators. I won't take your time to discuss that now, but that, too, presents a serious contract problem and a real cushion problem. I further suggest that you cannot discuss the contract problem without also discussing federal controls on the allocation of gas. I will name just a few: the abandonment requirements, the right-of-first-refusal requirements, and the geographical territorial protection, such as limited access to the outer continental shelf or other federal restrictions that limit access to markets and among markets. Some people are permitted to compete in certain areas for gas, and others are told they need not apply. They will not be eligible for competition regardless of what they're willing to pay. These subsets, these distortions, also need to be taken into consideration.

41

BENJAMIN ZYCHER: I have only one minor disagreement—with Russell's argument that it is difficult to make statements about the effect of total decontrol on consumers as a class. It seems to me that an improvement in the productive efficiency of the economy must make consumers as a class better off. It is difficult for me to conceive of a world in which the production of more goods makes consumers worse off.

Only two minor points I would like to make about a couple of the comments Ben Schlesinger made. First, the argument that the market is clearing seems to me to be a bit strange. My information indicates that we are looking at a world in which a good deal of nonprice allocation is going on. There are curtailment clauses in contracts, there are hookup bans here and there, and all sorts of nonprice allocated mechanisms are in force. I find it difficult to argue that the market is clearing in a world in which that sort of behavior is observed.

The last point I would make is that drilling statistics for January 1982 have fallen compared with those for January 1981. Under full decontrol of oil, world drilling is up 55 percent over that one-year period, while gas drilling under the provisions of the NGPA is down 1 percent.

EDWARD GRENIER: I think Gordon Gooch's point on the differences among industrial users is extremely well taken. In most of the economic analyses, industry tends to be put into one monolithic lump, and the model typically is the lower-priority bulk heat market. The issues on rate design that we will face as we move into the decontrol era more fully will be extremely complex. Although it is true that the feedstock process users will be part of the hostage group, I would put in a caution. There are some processes, which at a certain relationship between natural gas prices and electricity costs—namely, about a three-to-one ratio—can be switched to electricity. Once they are switched to electricity, they are gone forever from the gas system. But it will be in the best interests of all the higher-priority consumers, including the industrial higher-priority consumers, to keep a good bit of that lower-priority load on the system, provided that the lower-priority load is absorbing some of the fixed costs of the distribution in the pipeline systems. It will, therefore, be a very complicated question. I wonder if in the real world the market-clearing level will be much more complex. There may be several market-clearing levels. Pure economic analysis is fine, but it does not always fully reflect the real world. We have hard evidence that at least one pipeline has altered its purchasing practices because of the purchased-gas-adjustment litigation at the Federal Energy Regulatory Commission, notwithstanding the policy statement. That is all to the good because the prices that have been coming out of the section 107 gas are way out of line. That is another factor that has to be considered.

Gordon Gooch has raised some interesting questions on other subsets. Some of the questions, such as abandonment authority and right of first refusal, are issues that will stay with us and, again, may not be amenable to pure economic analysis. The question I want to ask is, What does the term "short run" mean? I know Ed Erickson said that the contract problems will go away in a year. I am a bit skeptical about that. I do not think we can sort these problems. I wonder if Dr. Russell has a view on that.

DR. RUSSELL: The short answer to the "short run" is that I do not know if "short run" refers to the timing of contract problems. I certainly take to heart many of Ed Erickson's comments on the question of what the costs are of trying to do something about the contract problems. Those have to be balanced against any disorder that is avoided by trying. I would be diffident, indeed, in suggesting any of the legislative remedies that have been suggested thus far along those lines.

Mr. Gooch, the point you made on the dual-fired question is very good. I thought I had dealt with the question of those users who are not bulk heat market users. They are among the inframarginal consumers. When approximately 30 percent of the total gas is being used in a market that is essentially bulk heat and approximately 15 percent of the gas is used in a market that has dual-fired capacity—that is, not 15 percent of the gas, but a market the size of 15 percent of the gas is shared by oil and gas in different proportions according to those prices—the point at which the prices are set and the market clears is in that marginal area. The inframarginal consumers will be the ones who will be faced with paying the higher prices, as long as the pipelines stay full in selling into the bulk heat markets.

JACK EARNEST: I will admit to some prejudice. As a lawyer I participated in the *Phillips (Phillips Petroleum Co. v. Wisconsin)* decision. For ten years or more, I was directly involved in selling gas. I have only recently become involved in buying gas. With that as background, I want to respond to Erickson on a couple of points.

I wish that pipeline management could take credit for being sagacious and wise and imaginative in its ways of trying to acquire gas supplies through promises and so forth. I have to admit, however, that when supplies are short, the imagination of the sellers probably contributed much more than that of pipeline management. I would add also that pipeline management is the gas supply guy sitting down in the field doing the renegotiation. It is also the gas buyer doing the gas buying. I think all pipeline managements are seriously concerned about the situation we have today. If there are any that are not, I have not met them. We do have some severe problems. We used to say that there are two ways to solve a supply-demand problem: one is to increase the supply, the other is to decrease the demand. I am not an economist, but that makes a certain amount of sense to me. The NGPA has been more effective in reducing demand than it has been in increasing supply, although there obviously have been some deliverability responses from the NGPA. In my judgment,

43

which is rather unfounded, I do not see us ever finding enough domestic gas to meet whatever the demand is. Demand can be fixed by price. Under the NGPA, we will fix a demand level that will greatly increase our reserve production ratio, so that we may have solved our problem. But assuming that we will continue to shoot for approximately eighteen to twenty-three trillion cubic feet of gas a year, I am convinced that we will not find that much gas over a period of time. It has been a long time since we found anywhere near that; maybe we did in 1981—I have not seen the numbers yet. Even so, I do not see us maintaining that level.

I would like to clarify a couple of points in deference to Schlesinger. The American Gas Association does speak for the pipelines. I think everybody should keep that in mind. People sometimes have a misconception that pipelines make a profit. When we talk in economic terms, people sometimes read one word and apply it to different situations. In case anyone does not know it, a pipeline makes absolutely zero profit on gas bought and sold; there is no margin. The pipeline buys it and resells it for exactly the price it paid. No one should have the misconception that we broker gas.

I want to make another point as we talk about the short run and renegotiating. Again, with deference to Erickson, it is my considered judgment from having sat on both sides for an extended time that the probability of renegotiating contracts in any so-called short run is zero. I have discussed this interminably with certain people—with many of you here at the table—and I know other people do not share my opinion. Most of the people I have contact with among the pipelines who do the negotiating and the renegotiating, however, probably feel more strongly about it than I do. I refer anyone who wants to consider how we renegotiate to some of the discussions that have been taking place for two years over the liquids and liquefiable issues; that will convince you how adequately we will be able to renegotiate on major contracts issues, which these are.

We have some strong institutional barriers to the operation of the gas industry. We have the institutional barriers of the state regulatory commissions trying to come up with realistic rate designs. This problem will take some time to solve. There are institutional barriers particularly in the contracts problem. The Interstate Natural Gas Association has been convinced, though not 100 percent so, and Texas Eastern certainly is convinced that specific changes are needed in the NGPA. We needed them last year, and we cannot afford to wait to see if Congress, before the general election, will come up with adequate answers to solve all of these problems. We could detail hundreds more problems than those we have already laid on the table. We will not solve all of those problems in any one format. The NGPA certainly does not solve them. We will have to come to an admittedly simple solution that will solve most of these problems and some of the significant problems we have under the existing NGPA.

ROBERT MEANS: If one must be concerned with investing money, then the question whether natural gas prices clear at 70 or 60 or 50 percent of crude is very important. If one views these matters as they properly should be viewed, as abstract policy choices, then it is not nearly so important where the market clears as whether it will clear, because on that issue turn two alternative and inconsistent arguments in favor of partial rather than complete decontrol. If one believes that the market will not clear, the consumers will receive in a direct way the benefit of the controls on a part of the supply of natural gas. If one believes that the market will clear, then price controls will subsidize the production of the categories of natural gas that are decontrolled. We can have one argument or the other, but we cannot have both arguments for partial decontrol.

In particular, I would like to raise a question about the equilibrating mechanism suggested by Russell when pipelines initially, as a result of the uneven distribution of the cushion, are confronted with varying purchased-gas costs. Russell's argument, as I understand it, is that there will be a market shift—on that point I agree—and that the new equilibrium would be reached because, in effect, as more decontrolled gas is added to the decushioned pipelines, the cushion is diluted. That equilibrating mechanism will not work for those pipelines that have a relatively small cushion, for those, in particular, whose cushion is smaller than the aggregate of their fixed costs, and for those of the distribution company customers because, for those pipeline systems as a whole, loss of market will lead to higher costs rather than to lower costs. In the end, one reaches what I call the black hole scenario in which they spiral into zero, and no light emerges from the system. To get equilibrium for those, one must abandon Russell's assumption, as I understand it, that a relatively fixed market-clearing price will result from essentially infinitely elastic demand at the margin, so that the market price in each market stays constant over wide variations in supply. One must assume that in the shallow cushion, the pipeline markets' equilibrium will be reached perhaps in the way Russell suggests. But it will also be reached partly as the market is forced to draw back to higher-valued uses that will support those high prices, which lead then to further inefficiency—simply the reverse of the inefficiency that existed under the Natural Gas Act when higher-valued uses in the interstate market were not served because you could not have gotten the gas. In the evenly distributed cushion, the higher-valued uses served by shallow cushioned pipelines would go unserved, but lower-valued uses served by deep cushioned pipelines would be served.

MARK COOPER: I have two general comments. First, I always like to say that intelligent people disagree only on their assumptions. I have waited to the end to see if I could hear all the possible assumptions before I had to jump in, and, fortunately, I have. We have heard every possible assumption about what competition will do to each of the theoretical problems that exist in the Natural Gas Policy Act. Competition created the contracts problem; competition will make it go away. A lack of competition will not make it go away. Competition creates supply; competition does not create supply. There is no competition, and so forth. The reason Ed Erickson has waited for what looked like a solution is that it is at the level of assumption that we differ, rightly or wrongly. Policy makers move very slowly when it is not simply different information but different assumptions that they are getting. I can hold back on giving my assumptions until later and not bear the cost of having displayed them.

Second, Russell's point on the partial versus general equilibrium analysis, that more gas in the partial equilibrium or more gas at a cheaper price, at a cheaper resource cost even in the partial equilibrium, is not necessarily more output in the general equilibrium. We can pay too much for it, or we can have too much of a disruption in order to get it in the general economy. That is an important point.

Another point is distribution. Even if we do get more output in the general economy, the distribution of output is very important. That is to say simply that if we have more goods and services in the society, we, therefore, must be better off. I do not accept that prima facie. The question is, Who has the goods and services? Great damage can be done to the distribution of wealth in society; we can have a bigger pie but also have much more skewed distribution. I do not accept that situation as better. I will have more to say about partial versus general equilibrium outcomes later on.

MICHAEL CANES: I, too, was very impressed with Russell's paper. It was an excellent laying out of the various issues showing where we are today. I would fault it only in that we do not go far enough simply to say that though the efficiency consequences are different for NGPA and full decontrol, it is not possible to measure the magnitude of these differences. I think it is extremely important for economists who look at this issue to try to provide some estimates of the magnitude of these efficiency consequences, and, furthermore, to explain to policy makers, and to people who work for policy makers, just how those get distributed. In that, I find myself agreeing with Cooper. It is much more interesting and much more relevant than economists sometimes seem to give credit for. People do care about who gains and who loses, and I think it is possible to identify some of these gainers and losers. I agree with Ben Zycher's comment that if, indeed, there are efficiency consequences, it does imply that a greater and more valuable total amount of goods and services will be available. It must mean that consumers are made better off. I would call for the economists who are looking at this problem to pay very close attention to that and to report their findings to people in Washington.

ELI BERGMAN, Americans for Energy Independence: I have noticed during this debate on the pace of decontrol that the parties to it have all given us their version of the effect of their favorite decontrol scenario on oil usage, whether it is more rapid decontrol, the same pace of decontrol, or slower decontrol. Some of you have waltzed around the issue, and I would like to know whether any of you would address more directly the effect on oil usage of the scenario you recommend.

MR. SCHLESINGER: Thanks for injecting that note. Unfortunately, there are many who do not worry themselves about whether foreign oil costs more or less. As long as it costs less, there presumably should not be any particular rule or regulation or, heaven forbid, a subsidy to prevent it from happening. That is a personal note. Eli, if the transient effect takes place, oil imports will increase. It is that simple, because the competition, as we have all been saying nine times over here, takes place in the boiler market, at least in the short term. In the longer term, my deep concerns are about what Grenier is suggesting, that the competition may switch to some other uses for gas. That is a severe concern. We recognize that aspect of it, but we are hopeful that there are so many inefficient uses of energy in this country that our ability to market gas will help alleviate some of those. One such inefficiency is that heat pumps are now capturing 50 percent of the home heating market, which is a joke given the other kinds of furnaces available today. Oil imports will rise if this transient effect takes the higher end, and we suspect that it will take a couple of years.

DR. RUSSELL: As the amount of discussion surrounding it today suggests, this gas supply question is highly complex. Let me make a few general comments and try to pull these things together. First, under Bob Means's black hole scenario for the gas pipeline companies, I interpret "black hole" to mean pipelines going through renegotiation. The pipelines, after the stockholders take a bath, will continue to be moving gas, under different management, perhaps, but certainly under different stockholders. It seems to me that, except as a transitory phenomenon, gas will find a market at the residual price or something less than the residual price. Consequently, I do not see either partial or full decontrol leaving gas unutilized in the field markets, except under very special and relatively short-term circumstances. The market will arrive for it.

The time path of what happens to the total gas supply is very important. Under full decontrol, there would be some incentives for in-fill drilling and for other actions affecting already existing and flowing gas, which will create a short-term spurt of gas. It certainly does mean that there is some elasticity in the supply of old gas, such that some gas will not be abandoned in reservoirs, some pumps will be maintained, part of the gas supply will be increased, and, more important, some of the old gas supply will move from the future back to the present. In looking at the time path of gas supply, I would suggest that more gas may well be found early on under full decontrol than under partial decontrol, which requires more time at the margin for exploration. But that does not tell us anything about what is really important, which is a couple of years out,

47

the intermediate period in which I would argue that more gas supplies will be brought on the market with partial decontrol than with full decontrol. The question is whether those additional gas supplies are worthwhile.

Bergman's question about oil imports and the view that backing out of oil imports is good in itself depends on two other questions. First, it depends on whether oil imports are the problem or oil vulnerability is the problem. Should we, on the one hand, try to minimize oil imports, or, on the other, try to deal with the problems of being vulnerable to oil insecurity? The second question is what it is worth to reduce that vulnerability in the event that reducing oil imports is a desirable way to go. It seems to me that it is clearly worth something, and I agree with Schlesinger that it is worth more than we pay for oil to reduce our oil imports. There probably is an oil import premium, but the question is how much it is worth and whether, if it is worthwhile to reduce oil imports, subsidizing high-cost gas in the field markets is an appropriate way to do it. These are my views on supply and oil imports and on the desirability of subsidizing through partial decontrol.

ROBERT WOODY: I would like to mention two issues. One, the price fight and the contracts problem fundamentally understate what we believe to be the fact, which is that the resource base is enormous. One can accept the potential gas committees' figures, and one can accept AGA figures, both of which we at GHK companies think are on the low side. We believe that from the Rocky Mountain overthrust to the eastern overthrust, the Michigan Appalachian base and the Anadarko, where only 4 percent of the sediments have been explored, domestically and internationally, methane is tremendously abundant. If that is true, and if it is also true that we now have a surplus of two trillion cubic feet in the market, one could conclude that there will not be a price fight. As I understand Russell, he more or less agrees with that in the macro sense. He used the word "discontinuity," I believe, with respect to regional situations. We believe that is absolutely the case, and Merrill Lynch in its January report states that in its opinion there will be no price fight in 1985.

Second, as to the contracts problem, with all deference to my friend Mr. Earnest, producers cannot store it, they cannot eat it, they have to sell gas; and that being the case, with bills due at the bank, there will be renegotiation. We believe that before you invoke the power and majesty of the United States government to abrogate contract terms, it should be determined whether that is a real or a hypothetical problem. We believe it is a hypothetical problem. The revenue stream touches on the fundamental part of the argument, at least as we see it. The revenue goes to new gas producers, those who are betting on the gas that is yet to be found, those who are not seeking to be paid for gas found long ago when it was a byproduct of the search for oil. Economics before the 1978 NGPA were quite satisfactory. We believe that is the heart of the argument and that, moreover, a market shares battle may be occurring. It may be that the major oil companies would rather sell residual oil than natural gas and that the nuclear industry and the coal industry may believe, as we believe, that a lot of natural gas is out there. It may not be foolish to think that it can be used for cogeneration and for methonal purposes and so forth. We believe that the debate is over old gas versus new gas, and if we try, through hypothetically posed problems, to resolve it now, we invite not controls that will phase out over time but a permanent system of controls in the form of a windfall profits tax. It is clear now that this tax on crude oil is permanent.

EDWARD MITCHELL: I think that is an appropriate place to stop for this session. I will make one observation: I expect that, at these gatherings, what to an economist is an adjustment to equilibrium is to another man, change of management or bankruptcy.

Part
Two

Pitfalls on the Road to Decontrol: Lessons from the Natural Gas Policy Act of 1978

Catherine Good Abbott and Stephen A. Watson

A primary goal of the Natural Gas Policy Act of 1978 (NGPA) was to ensure a smooth transition to partial decontrol in 1985. In the three and a half years since its passage, three key problems with the NPGA have been identified and debated. First, the sudden increases in world oil prices since 1978 appeared to render the price trajectories in the NGPA obsolete, creating the possibility of a large gap between regulated gas prices in 1984 and partially deregulated gas prices in 1985. Second, because deregulation in 1985 will be only partial, the uneven distribution of price-controlled gas after 1985 (the "price cushion") among pipelines and markets may seriously distort the allocation of gas supplies to end users. Finally, pipelines are concerned that contracts signed during long periods of price regulation may cause gas prices to overshoot market-clearing levels, making their gas unmarketable.

Although this paper will focus on the two market-ordering issues—the uneven distribution of the price cushion and the so-called contracts problem—the potential for a price gap is sufficiently important to warrant a brief review.

Pricing Problems under the Natural Gas Policy Act

Congress enacted the NGPA in 1978. The act brought intrastate gas under federal regulation for the first time and imposed complex price controls on gas production. There are more than twenty categories for natural gas pricing, but they can be simplified into four groups:

1. "new" (post-1977) gas, most of which receives a 3–4 percent real price increase annually, reaching the equivalent of $15 per barrel (in 1978 dollars)

The views expressed in this paper are entirely those of the authors and do not necessarily represent official positions of the Department of Energy or the Interstate Natural Gas Association of America.

by 1985, when most "new" gas may be decontrolled

2. "old" (pre-1977) *interstate* gas, which remains price controlled at 1978 real prices until exhausted

3. "old" (pre-1977) *intrastate* gas, of which some is deregulated in 1985 and the price-controlled remainder receives a regulated price higher than that of old interstate gas

4. certain high-cost gas (wells deeper than 15,000 feet and unconventional gas other than tight sands gas), which is deregulated (most of this gas, less than 5 percent of today's supplies, currently sells at prices well above "free-market" levels)

The NGPA pricing provisions will leave 40 to 50 percent of domestic gas supplies under price controls in 1985 and approximately 20 percent in 1990.

A major conclusion of the exhaustive review of natural gas pricing policies by the Department of Energy (DOE) was that the NGPA is likely to leave a large price gap between 1984 controlled prices and 1985 partially decontrolled prices.[1] Under world oil price assumptions then current,[2] average wellhead prices were projected to increase between 1984 and 1985 by 70 percent (above the rate of inflation), and residential prices were projected to increase by 36 percent.

The price gap depends on three critical assumptions:

- the projected world oil price in 1985
- the projected natural gas wellhead price in 1984
- the projected relationship between crude oil prices and wellhead gas prices upon partial decontrol in 1985

The recent softening in world oil markets has led some observers to question whether a price gap will in fact be a problem in 1985. In addition, there has been considerable debate over the likely future relationship between market-clearing levels for average gas prices at the wellhead and crude oil prices. When discussing full decontrol, pipelines tend to argue that wellhead prices for gas (including severance taxes) will range from 50 to 70 percent of crude oil acquisition costs of refiners. Producers have argued that the relationship is likely to be 80–90 percent of crude oil costs. Energy Action assumes that wellhead gas prices will rise to 100 percent of crude oil prices.

Finally projections of average domestic gas prices in 1984 range from $2.31[3] to $3.44 per thousand cubic feet[4] (in 1980 dollars). The American Gas Association argues that "category creep," the movement of gas from lower-priced categories to higher-priced categories, accounts for their higher estimate. All the major published estimates, however, include increased production from developmental wells (a major component of category creep).

Other analysts have asserted that the price gap is likely to be much smaller than the DOE study projects because the world oil markets are likely to be

TABLE 1
PROJECTIONS OF GAS PRICE INCREASES, 1984–1985

	DOE[a]	1981 AGA[b]	Merrill Lynch[c]	1982 AGA[d]
Wellhead price increase, 1984–1985 (percent)[e]	70	7[f]	(1–2)	23[f]
Wellhead gas price, 1984 (1980$/thousand cubic feet)	2.61	3.44	2.65–2.83	2.46
Gas price as percentage of crude oil price, 1985[g]	66	57	65[f]	61[f]
Crude oil price, 1985 (1980$/barrel)	35.00	34.21	23–24[f]	26.50
Annual wellhead price increases, 1980–1984 (percent)[e]	10	21	15–17	9

a. DOE, *A Study of Alternatives to the Natural Gas Policy Act of 1978,* November 1981.
b. American Gas Association, "Consumer Impact of Indefinite Price Escalator Clauses under Alternative Decontrol Plans," November 6, 1981. We have taken the NGPA "optimistic" scenario, which assumes no contract problem, for the purposes of comparison. AGA's central case shows a 51 percent wellhead price increase.
c. Merrill Lynch, *Natural Gas Monthly: Outlook for 1982,* a Merrill Lynch industry review, January 1982. Merrill Lynch shows a 6–7 percent *nominal* increase, with an assumed 8 percent inflation rate, implying a *real* decline of 1–2 percent between 1984 and 1985.
d. American Gas Association, "The Spring A.G.A.–TERA Base Case," July 13, 1982. In this forecast AGA does not forecast prices under the NGPA if no contract problem develops. With a contract problem, AGA projects a 39 percent increase in average wellhead prices between 1984 and 1985. AGA states that its best estimate of a market-clearing price for deregulated gas would be 65 percent of crude oil prices, including taxes (see p. 9). We have used this forecast to estimate the 1985 wellhead price AGA might project if no contract problem developed.
e. Above inflation.
f. Estimated.
g. The gas price in this calculation excludes severance and other taxes (estimated at 7 percent) because the maximum lawful prices in the NGPA and in much of the proposed legislation exclude such taxes. The ratio increases by roughly 5 percent if taxes are included.

softer, leading to lower oil prices in 1985; natural gas prices are likely to be higher in 1984 because the gas cushion is disappearing faster than anticipated; and wellhead gas prices are likely to be lower relative to crude oil than the DOE projects. Table 1 compares projections of oil and gas prices by three organizations: the DOE, the American Gas Association (AGA), and Merrill Lynch. Two estimates are shown for AGA, their late 1981 forecast (the most recent available for the presentation of this paper) and their spring 1982 forecast (the most recent when the paper was being revised).

Both the Merrill Lynch and the 1981 AGA forecasts project a smooth transition to decontrol in 1985, but for different reasons. The 1981 AGA forecast projects a higher wellhead gas price in 1984 than DOE under the NGPA (by roughly $0.80 per thousand cubic feet) and a lower relationship of gas prices to crude oil prices in 1985.[5] The price gap does not develop in

AGA's 1981 forecast because AGA projects a 21 percent real annual increase in wellhead prices to 1984, while DOE projects a 10 percent per year increase.

Merrill Lynch shows a range of 1984 prices similar to DOE's estimates and projects a similar ratio of crude oil to gas prices in 1985. Merrill Lynch, however, sees world oil prices staying flat in *nominal* dollars (and declining to the low to middle twenties in real 1980 dollars).

In July 1982, AGA revised its estimates of average wellhead prices in 1984 downward by almost $1.00 per thousand cubic feet, to a level just below DOE's forecast.[6] Thus, by the summer of 1982, the range of projected 1984 wellhead prices under the NGPA had narrowed substantially. Considerable debate continued, however, over the likely level of crude oil prices in 1985 and the relationship between gas and oil prices.

Table 2 and figure 1 explore the effect on the price gap under a range of assumptions about 1985 world oil prices and the relationship between average gas prices and crude oil prices under the NGPA. Because the range of projected 1984 wellhead prices has narrowed, table 2 and figure 1 use DOE's forecast as representative. Under DOE's forecasts, for example, wellhead prices will increase by 30 percent (above inflation) if world oil prices are above $30 per barrel and gas prices clear at more than 60 percent of crude oil prices.[7] Figure 1 illustrates these relationships by drawing isopercentage lines showing what combination of world oil prices and ratios of oil to gas prices produce 25 percent increases in gas prices between 1984 and 1985.

From this analysis we can conclude that it is highly likely that the price gap will be a major problem if 1985 world oil prices rise above $25 per barrel ($30 in 1982 dollars) and gas prices clear above 65 percent of crude oil prices. Conversely, if world oil prices are lower than $25 per barrel and the gas-oil price ratio is lower than 65 percent, we would not anticipate a large average jump in gas prices with partial decontrol in 1985.

The disadvantage of the NGPA's rigid pricing formulas is their inability to respond to changes in gas and oil market conditions. The NGPA raises the prospect of a large price jump as a result of decontrol if world oil prices are high in 1985. If world oil prices are low, *average* gas prices may make a smooth transition to 1985 decontrol, but the market disorder problems discussed in the next section of this paper will remain.

Market Disorder Problems

The main topic of this paper is the potential for a disorderly gas market at the time of decontrol caused by two potential problems: the uneven endowments of cheap gas and the so-called contracts problem. The first problem stems from the fact that some regions and pipelines will have less access to cheap price-controlled gas after partial decontrol in 1985 than others. These uneven endowments of cheap gas may affect who is able to obtain new gas supplies in 1985.

TABLE 2

WELLHEAD PRICE INCREASE UNDER THE NGPA

BETWEEN 1984 AND 1985: DOE ASSUMPTIONS

(percent)

Assumed 1985 Crude Oil Price		Assumed Ratio between 1985 Average Domestic Wellhead Prices for Natural Gas and Crude Oil Prices[a]							
1980$/barrel	1982$/barrel	60	65	70	75	80	85	90	100
22.50	26.50	—	6	14	21	30	38	46	62
25.00	29.45	8	18	26	35	45	53	62	80
27.50	32.39	19	30	39	48	59	69	78	98
30.00	35.34	30	41	52	62	74	84	94	116
32.50	38.28	40	53	64	75	89	99	110	134
35.00	41.23	51	65	77	89	103	115	126	152
37.50	44.17	62	77	90	102	117	130	143	170
40.00	47.12	72	89	102	116	132	145	159	189

NOTE: DOE projects an average domestic wellhead price of $2.61 per thousand cubic feet (in 1980 dollars) under the NGPA in 1985. See *Study of Alternatives,* table 1, p. 9.

a. Wellhead prices are used here to mean the price paid by the pipeline to the producer less severance and other taxes. Legally, this would be the maximum lawful price, since severance taxes (assumed to average 7 percent) are allowed on top of maximum lawful prices. The ratios would be 64 percent, 70 percent, 75 percent, 80 percent, 86 percent, 91 percent, 96 percent, and 107 percent, respectively, if taxes were included.

FIGURE 1

WELLHEAD PRICE INCREASE UNDER THE NGPA

BETWEEN 1984 AND 1985: DOE ASSUMPTIONS

Assumed 1985 Crude Oil Price		Assumed Ratio between 1985 Average Domestic Wellhead Prices for Natural Gas and Crude Oil Prices							
1980$/barrel	1982$/barrel	60	65	70	75	80	85	90	100
22.50	26.50								
25.00	29.45								
27.50	32.39								
30.00	35.34								
32.50	38.28								
35.00	41.23								
37.50	44.17								
40.00	47.12								

NOTE: See footnotes to table 2.

The second problem, raised by pipelines and distribution companies, is that some pipelines have signed contracts that they believe may make their gas unmarketable upon partial decontrol in 1985. Both problems pose risks for an orderly transition of the gas market to partial decontrol in 1985.

Uneven Endowments of Old-Gas Supplies. The NGPA has different decontrol provisions for intrastate gas from those for interstate gas. Some old (pre-1977) intrastate gas will be decontrolled in 1985, and much of what remains price controlled will be eligible for a relatively high price. In contrast, *all* old interstate gas will remain price controlled (generally at lower prices than old intrastate gas), as will some new (post-1977) interstate gas.[8] This differential treatment of interstate and intrastate gas creates the potential for market instability at the time of decontrol. DOE estimates that 53 percent of interstate domestic gas supplies and only 43 percent of intrastate supplies will remain price controlled in 1985. More significant, the interstate market has a price advantage over the intrastate market in 1985 of $0.80 per thousand cubic feet for these old-gas supplies. These differences may create the potential for uneven bidding between markets for deregulated gas and may lead to unequal access of markets to new-gas supplies.

Moreover, price cushion disparities are substantial within each market. Within the intrastate market, much of the price cushion is owned by industries and electric utilities that buy their gas directly from producers. In 1978 intrastate pipelines paid an average price of $1.60 per million Btu while direct purchases by electric utilities averaged $0.95 per million Btu and direct industrial purchases averaged $0.71 per million Btu. Electric utility and industrial direct sales constituted 34 percent of the market in 1978. Yet 43 percent of the old gas still under contract to these direct purchasers in 1985 will be governed by definite escalator clauses at low prices, and only 23 percent of intrastate pipeline purchases of old gas in 1985 will be from these inexpensive sources.[9]

The price disparities among interstate pipelines today are quite substantial. Using data on purchased-gas costs filed at the Federal Energy Regulatory Commission (FERC) in late 1981 and early 1982, the Energy Information Administration (EIA) found that the average purchased-gas costs range from $1.30 to $2.76 per thousand cubic feet, a factor difference of 2.1.[10] Interstate pipelines also vary substantially in their access to old gas (sections 104 and 106), ranging from 29 percent to 79 percent.

Some observers have argued that the EIA data contradict the hypothesis that market disorder is likely to result from uneven endowments of cheap price-controlled gas. GHK Companies, for example, asserts the following: "Contrary to earlier statements, data published in December by the Energy Information Administration reveal no relationship—in fact, a slight inverse relationship—between the size of a pipeline's old gas 'cushion' and the amount the pipeline has agreed to pay for unregulated gas."[11] This assertion bears close

TABLE 3
COMPARISON OF INTERSTATE PIPELINES

Pipeline	% 107	% new	% old	Average price	1980 Average Price Form 2[b]	Ratio of 1973 Volume to 1977 Volume	Reserve Adds[c]
Cities Service	13	43	44	1.98	1.28	1.23	0.23
Colorado Interstate	13	40	46	2.65	1.92	1.05	0.31
Columbia Gas	9	53	38	2.74	1.83	1.69	0.33
Consolidated	0	71	29	1.99	2.27	1.06	0.47
El Paso	4	37	59	1.92	1.64	1.71	0.33
Florida Gas	6	28	66	1.88	1.55	1.16	0.28
Kansas-Nebraska	5	21	74	1.62	1.25	1.01	0.21
Michigan-Wisconsin	3	26	70	1.86	1.61	1.17	0.16
Natural Gas Pipleline Co. of America	0	22	78	1.80	1.56	1.03	0.40
Northern Natural	3	29	68	1.56	1.33	1.30	0.16
Northwest Pipeline	0	37	62	1.89	1.56	n.a.	0.18
Panhandle Eastern	2	34	64	1.64	1.36	1.26	0.29
Southern Natural	14	30	56	2.67	1.69	1.27	0.37
Tennessee Gas Pipeline	2	27	71	2.04	1.52	1.18	0.28
Texas Eastern	0	20	79	1.30	1.00	1.06	0.31
Texas Gas	8	22	70	2.17	1.49	1.39	0.28
Transcontinental	10	47	43	2.76	2.07	1.73	0.55
Transwestern	4	48	48	1.96	1.46	1.41	0.41
Trunkline	9	23	68	2.53	1.75	1.09	0.75
United Gas Pipeline	9	56	34	2.35	1.87	1.30	0.69

n.a. = not available.
a. Purchased-gas adjustment data from EIA, *The Current State of the Natural Gas Market,* December 1981, p. 66.
b. Wellhead purchases (accounts 800, 801, 802).
c. Gross changes in reserves from 1977 through 1979 divided by reserves remaining on December 31, 1979.

inspection because the EIA data referred to provide an excellent opportunity to investigate how the gas market is functioning today.

Seven interstate pipelines now purchase 9 percent or more of their gas supplies from high-cost sources (section 107). While much of section 107 gas is deregulated, tight sands gas is regulated at a relatively high price ($5.22 per thousand cubic feet). Although it is not technically precise to do so, section 107 purchases are deemed deregulated for the purposes of this discussion. The *average* price for this gas ranges from $4.02 to $8.04, with five of the pipelines paying average prices of $7–8 per thousand cubic feet. Most observers would agree that pipelines could not pay $7–8 for deregulated gas if there were no cheap price-controlled gas to "roll in" with the high-cost gas. Thus the current data support the hypothesis that pipelines will tend to use their price cushion to obtain deregulated gas supplies.

It should be noted that the concept of a pipeline's cushion is more complicated than the quantity of old gas purchased by the pipeline at the wellhead. First, pipelines buy from other pipelines. At first glance, Consolidated appears to be in a very poor bidding position with only 29 percent of its gas old (see table 3). Consolidated purchases large quantities of gas, however, from Tennessee Gas Pipeline, which has 71 percent old gas; Texas Eastern, with 79 percent; and Texas Gas, with 70 percent. Thus the interconnections among gas pipelines must be accounted for in measuring the cushion. Second, the price cushion comprises all gas that remains price controlled after 1985, not just old gas. Significant quantities of new gas—section 102(d) gas and section 103 gas that was dedicated to interstate commerce before the NGPA, as well as sections 108 and 109 gas—remain price controlled after 1985 (or 1987). The EIA data do not distinguish between "forever controlled" and "to be decontrolled" gas pipeline by pipeline. Although old gas is the largest component of the cushion, it is not a fully adequate measure of the size of the cushion. Finally, imported gas from Canada and Mexico (in addition to domestically deregulated gas) also "uses up" the cushion. A thorough analysis of current bidding behavior would take into account these complexities.[12] The rest of this paper will use the simplifying assumption that the percentage of old gas is indicative of the size of a pipeline's cushion.

GHK poses an interesting question: Why is it that the pipelines that are most aggressively purchasing deregulated gas do not appear to be the pipelines with the largest cushion? Several hypotheses come to mind:

• Pipelines will bid for high-cost deregulated gas if they need the gas to meet short-term delivery requirements. Pipelines that do not need gas to meet existing load or cannot add new load in the short run are unlikely to bid for high-cost gas.

• Pipelines may bid for gas on the basis of their medium- to longer-term reserve positions. Pipelines with low reserve-to-production ratios may be willing to bid more for new gas than pipelines with a longer reserve life.

• Pipelines with larger cushions may be more successful than pipelines with smaller cushions in bidding for gas that is now regulated but will be deregulated in 1985. The larger-cushion pipelines may be able to offer higher prices upon partial decontrol in 1985 than pipelines with smaller cushions. Thus the small-cushion pipelines may be forced to use up their cushion now by bidding on already deregulated gas. Bidding high prices today for already deregulated gas will put the small-cushion pipelines in an even worse bidding position in 1985.

TABLE 4
CHANGES IN INTERSTATE PIPELINE SUPPLIES, 1973–1980
(billion cubic feet)

Pipeline	1980	1979	1978	1977	1975	1973
Cities Service	412	454	460	452	461	558
Colorado Interstate	409	421	426	429	438	449
Columbia Gas	1,121	1,147	1,099	1,041	1,126	1,764
Consolidated	638	700	655	650	627	690
El Paso	1,291	1,267	1,190	1,146	1,339	1,959
Florida Gas	186	173	139	119	94	138
Kansas-Nebraska	121	128	137	133	134	134
Michigan-Wisconsin	691	807	773	755	865	883
Natural Gas Pipeline						
Co. of America	1,016	1,109	1,041	1,030	1,060	1,058
Northern Natural	800	794	744	746	927	973
Northwest Pipeline	353	441	403	446	410	n.a.
Panhandle Eastern	732	738	688	633	704	799
Southern Natural	614	639	598	544	651	690
Tennessee Gas Pipeline	1,147	1,119	1,118	1,136	1,212	1,340
Texas Eastern	981	946	870	869	761	921
Texas Gas	420	638	577	567	670	790
Transcontinental	919	739	593	564	746	975
Transwestern	335	310	235	270	308	382
Trunkline	493	511	452	412	364	449
United Gas Pipeline	1,038	950	839	873	919	1,132
Total	13,717	14,031	13,037	12,815	13,816	16,084

SOURCE: FERC form 15.
n.a. = not available.
NOTE: Includes purchases from other pipelines.

Although a complete analysis of these alternative hypotheses is not yet available, several observations can be made. First, pipelines with high purchases of deregulated gas account for a large percentage of the increase in interstate purchases since 1977 (a major shortage year). The top twenty interstate pipelines increased their wellhead purchases by 900 billion cubic feet between 1977 and 1980. Eighty-three percent of the increase (750 billion cubic feet) can be attributed to five of the seven pipelines that purchase at least 9 percent deregulated gas (see table 4). Two pipelines alone (Transcontinental and United) account for 520 billion cubic feet of the increase. Two of the seven pipelines have had declining volumes since 1977.

Some pipelines that experienced significant load loss in the 1973 to 1977 period (as signified by a high ratio of 1973 to 1977 volumes) and have increased load since 1977 are not heavily engaged in 107 gas purchases (El Paso has only 4 percent 107 gas, Northern Natural has 0 percent, Panhandle Eastern 2 percent, and Transwestern 4 percent).[13] Several of the seven big deregulated gas purchasers appear to have avoided large curtailments in the 1973 to 1977 period: Colorado Interstate and Trunkline had relatively flat sales from 1973 to 1977. Thus a cursory examination of the data does not reveal a close relationship between load loss in the mid-1970s and active purchases of section 107 gas.

Second, pipelines particularly aggressive in their purchases of deregulated gas do not appear to be unusually successful in adding reserves. All seven of the big deregulated gas purchasers have replaced more than 23 percent of their reserves in the 1977–1979 period. Three of the seven pipelines (Transcontinental, United, and Trunkline) have replaced more than 55 percent of their reserves in that period. Nine other pipelines have also replaced more than 23 percent of their reserves in the same period; thus rapid increases in reserves are not confined to the large purchasers of 107 gas. Other pipelines appear to be more successful in adding reserves without needing to bid for high-cost deregulated gas.

Evidence on ability to obtain section 102 gas is not readily available. There appear to be no significant trends in the percentage of section 102 gas obtained by large- or small-cushion pipelines, but percentage of sales volume may not be a good measure of new-gas purchasing ability since some 102 gas was already contracted for before the NGPA.

In conclusion, further analysis of individual pipelines' behavior is necessary to sort out the differential effect today of uneven price cushions. The available evidence does, however, support the hypothesis that some pipelines are bidding more for deregulated gas than they would bid if they had no price cushion. Some analysts have argued that because pipelines today are not "using up" all of their price cushion in bidding for deregulated gas, a majority of the cushion will not be used up in 1985. This argument ignores the fact that less than 5 percent of today's domestic gas supplies are deregulated and more than 50 percent will be deregulated in 1985. Much of the gas deregulated in 1985 is already under contract. The pricing provisions governing these contracts (discussed in the next section) indicate that pipelines have in fact been bidding more for this gas than they could resell it for at the margin.

The "Contracts" Problem. Some elements of the natural gas industry argue that existing contracts could cause a disorderly market at the time of decontrol and that the potential for disruption is sufficiently high to warrant federal intervention in existing contracts. Concern about this problem has motivated both the American Gas Association (AGA) and the Interstate Natural Gas

Association of America (INGAA) to support reconsidering the NGPA. Distribution companies and state public utility commissions have also expressed concern. These groups argue that the contracts problem is at least as important a problem as price controls on old gas, if not more important.

The potential problem could be caused by the joint operation of two types of contracts:

• *Oil-tied contracts,* which tie the price of gas upon deregulation to the Btu equivalent of an oil product price (for example, distillate) at the wellhead, generally with no allowance for transportation to end-use markets

• *Indefinite price escalator contracts,* which tie the price of gas to other gas contracts in an area (for example, a most-favored-nation clause, which bases the price on the average of the three highest prices in a region)

A relatively small number of oil-tied contracts could trigger a large number of most-favored-nation contracts to prices well above market-clearing levels (for example, the price of distillate fuel oil). Pipelines are concerned that their *average* price would rise above market-clearing levels, leading to substantial loss in gas sales and increases in oil use. While pipelines can attempt to minimize this load loss by reducing their purchases of decontrolled gas, they believe that the take-or-pay minimums in their existing contracts would severely constrain their ability to reduce average gas costs quickly. The AGA estimates that the gas industry could experience a loss in sales of 1.4 to 2.5 trillion cubic feet if the problem is not addressed.[14]

The AGA contends that "the NGPA will provide a smooth transition from 25 years of price controls to free market prices, so long as indefinite price escalator clauses are legislatively defused."[15] Thus some elements of the gas industry have identified the contract issue as *the* key problem with the NGPA.

In contrast, analysts at Merrill Lynch have concluded that there is no contracts problem:

We do not believe that IDPE or MFN [indefinite-price-escalation or most-favored-nation] clauses are going to create the problems portended by the industry. Producers could ultimately agree to settle for prices well below the cost of No. 2 fuel oil, otherwise pipelines may not be able to market the gas. As a result, a significant gas surplus could develop. Pipelines would not be able to take all of the gas they had contracted for and we doubt that they would honor the take-or-pay provisions contained in their contracts. That could ultimately force producers to settle for lower prices.[16]

Two basic questions are at stake:

• *The fact question:* How many contracts of each type will be in effect in 1985?

- *The behavioral question:* How quickly will the private sector renegotiate those contracts if the problem develops?

The fact question can be answered at least with regard to today's contracts. The behavioral question is the major area of dispute: the pipelines argue that allowing the private sector to handle the problem will be a long and messy process lasting more than three years. Those opposed to intervention argue that the private sector is better equipped to handle the issue and that the renegotiation process could be completed much more quickly.

The Size of the Potential Problem. The publicly available data base on contracts was limited at the time this paper was presented. Interstate contracts on file at FERC include all pre-NGPA and some post-NGPA contracts. Intrastate contracts were surveyed in 1978, and the results of the surveys were presented at the conference. Since then, however, results of the Energy Information Administration's random survey of post-NGPA contracts have become available.[17] While the new survey results do not change the thrust of our arguments, the data base is considerably stronger. Therefore, the rest of this section relies on the results of that survey.

On the basis of EIA's data, the following observations can be made (see table 5):

- Oil parity clauses (without market-outs, which free the buyer from contractual obligations if the gas is not marketable) are very infrequent (roughly 1 percent) in contracts for gas to be deregulated under the NGPA. Oil parity terms are more prevalent in all gas volumes.
- Oil parity clauses are not concentrated either by ownership or by geography, thereby increasing the likelihood of contagion.
- Most-favored-nation clauses (without market-outs) are included in roughly 55 percent of interstate contracts deregulated under the NGPA but in less than 15 percent of intrastate contracts. These contracts cover almost 30 percent of the total volume of gas deregulated in 1985. Thus most-favored-nation clauses, if triggered by oil parity contracts and not renegotiated, could pose a significant transition problem for the gas market under the NGPA.
- A large percentage of deregulation clauses (or indefinite price escalators) are not specifically most-favored-nation clauses. Deregulation clauses cover 51 percent of gas deregulated under the NGPA, and most-favored-nation clauses cover only 33 percent. Instead of a most-favored-nation clause, many contracts provide for open-ended yearly negotiations of fair market value or for inflation increases from a price floor upon decontrol.
- Market-out provisions are relatively frequent among NGPA-deregulated interstate contracts. Two pipelines, Transcontinental and Michigan-Wisconsin, have already used market-out provisions to reduce the price of their high-cost deregulated gas.

TABLE 5

GAS VOLUMES TO BE DEREGULATED IN 1985 UNDER THE NGPA

(percent)

	Deregulation Clause	Deregulation Clause with Most-Favored-Nation Clause	Deregulation Clause with Most-Favored-Nation Pro-vision and No Outs	Oil Parity Clause with No Outs
Interstate	86	72	56	2
Intrastate	35	14	13	—
Total	51	33	27	1

SOURCE: Based on EIA, *Natural Gas Producer/Purchaser Contracts,* tables 15–17.

In contrast, both the AGA[18] and INGAA[19] have argued that as much as 80 percent of gas deregulated in 1985 will be subject to the "contract fly-up." These estimates are consistent with EIA's finding that 86 percent of all inter-state volumes deregulated under the NGPA have deregulation clauses. The AGA and INGAA may be overstating the problem, however, by treating all deregulation provisions as if they were most-favored-nation clauses with no market-out or other buyer protection clause.

In addition, the AGA and INGAA may draw their data predominantly from the *interstate* market. As table 5 shows, EIA found a significant differ-ence in contracting patterns between the interstate and intrastate markets. Industry analysts have questioned the low frequency of most-favored-nation clauses in EIA's intrastate sample, particularly for intrastate pipelines. Thus the EIA data are not universally accepted as representative of intrastate con-tracting practices.

The small percentage of oil parity clauses (without market-outs) in EIA's survey raises the question whether oil parity clauses are sufficiently frequent to "infect" most-favored-nation clauses. EIA argues that there is a "critical mass" of oil parity contracts, largely because of their wide geographic dispersal. Many most-favored-nation clauses, however, refer to an *average* of three contracts with similar contract terms. In addition, a third of EIA's oil parity sample use some kind of adjustment to subtract the costs of transportation to the burner tip (a "transportation netback"). These netbacks lower the wellhead prices in the oil parity contract. Of course, oil parity clauses may not be the highest-price term in a contract. Some contracts reportedly have a high fixed-price term (for example, $8 per million Btu) plus an inflation escalator. EIA does not present data on this type of price floor.

The best survey of today's contracting practices cannot reveal what mixture of contracts will govern decontrolled gas in 1985. The behavioral questions dominate the debate about the potential for contract fly-up in 1985:

• Will new contracts signed between now and 1985 give buyers sufficient flexibility to avoid the contracts problem?

• Will buyers have sufficient leverage between now and 1985 to force renegotiation of existing high-priced or problem contracts?

• If problem contracts still exist in 1985, how quickly will the contract infection spread, and how wide will the contagion be?

• If average gas prices rise above market-clearing levels, how quickly will the ensuing industrial load loss discipline the market?

• To what extent will continued price controls on some gas under the NGPA and deregulated gas subject to low contractual prices mitigate the effects of contract fly-up? Will pipelines take their contract problem into account when bidding for deregulated gas in 1985?

Uncertainty about the answers to these behavioral questions has prompted substantial interest in a government-imposed solution to the contracts problem.

Policy Considerations. By imposing price controls, the federal government prevented interstate pipelines from negotiating with producers on price and encouraged pipelines to compete for gas through contract terms. The partial deregulation strategy in the NGPA encouraged pipelines to sign high-priced (for example, oil parity) contracts that anticipated use of the gas cushion to "roll in" high-cost gas with cheaper price-controlled gas.

Indefinite price escalator clauses are not per se a product of regulation. One survey shows that by 1978 half of the unregulated intrastate market was covered by such contracts.[20]

There are two major policy questions:

• Does the risk of the disruptive period of contract renegotiation justify federal intervention in contracts?

• Can legislative intervention be designed that would help, not hinder, the transition process?

Policy Options. Five types of policy options have been proposed as solutions to the contracts problem:

Option 1: do nothing
Option 2: the INGAA approach
Option 3: the prohibition strategy
Option 4: the containment strategy
Option 5: the Gramm proposal

Each option is considered in turn below.[21]

Option 1: do nothing. The "do nothing" strategy assumes that pipelines and producers will have sufficient opportunity and incentive to readjust onerous contract terms to reflect the market realities of partial decontrol.

If average prices rise above market-clearing levels and significant load loss occurs, producers are likely to start cutting prices in order to market their gas. This price reduction is likely to occur because the wellhead market is "workably competitive." The key issue is how long the price renegotiation process takes. The concern is that it will take sufficiently long that significant market disorder may occur.

Pipelines may be able to offer producers incentives to renegotiate high-priced contracts by agreeing to renegotiate low-priced contracts. Not all pipelines, however, will have both types of contracts with all producers. Moreover, producers' flexibility may be somewhat constrained by legal obligations to royalty owners.

Option 2: the INGAA approach. INGAA has proposed an option to cap all gas under contract on date of enactment at 70 percent of the crude oil price (*including* severance taxes) until the contract expires or is renegotiated.[22] In effect, the proposal would set the price for 65–80 percent of all interstate (and a majority of intrastate) volumes, even after price controls formally ended. Already deregulated gas (section 107 gas) would be frozen at current contract prices until the 70 percent cap sets a higher price. Renegotiated contracts and new contracts would not be subject to the cap.

The proposal's basic flaw is that it relies heavily on the federal government, rather than the market, to establish gas prices close to market-clearing levels: this cap could set prices for large volumes of gas well into the 1990s. If the market-clearing price for gas is above 70 percent of the crude oil price, the cap would effectively re-regulate some gas at 70 percent. If the gas market clears below 70 percent of the crude oil price, the cap could act as a floor, rather than a ceiling, and keep prices artificially high. The option does provide, however, for relatively easy escape from the cap if both parties agree to renegotiate (although if most buyers and sellers would renegotiate, the intervention is not needed in the first place).

Option 3: the prohibition strategy. The prohibition strategy prohibits certain types of contract clauses (primarily indefinite price escalation clauses) for both existing and new contracts and restricts the ability of pipelines to sign take-or-pay provisions in contracts. In effect, this option seeks to regulate the contract terms between pipeline and producer out of a fear that competitive pressures will not be sufficiently strong once wellhead price controls are removed. Advocates of this option often cite pipelines' behavior in purchasing already deregulated (section 107) gas as evidence of the need to impose restrictions on existing and future contracts.

While there is a legitimate argument that existing contracts may reflect unequal bargaining power between pipelines and producers, it is difficult to argue for federal controls over *future* contract terms if one accepts the proposition that the wellhead market is workably competitive. The question of the competitiveness of the wellhead market for natural gas was debated extensively in the 1970s and culminated in a definitive Federal Trade Commission staff study, which found that "the natural gas industry is capable of workably competitive performance in the absence of Federal price regulation."[23]

The prohibition strategy thus appears to be a mechanism for backdoor regulation of wellhead prices for natural gas.

Option 4: the containment strategy. The containment strategy seeks to limit the effect of ill-advised existing contract terms by treating all existing contracts—for the purposes of triggering any indefinite price escalation clauses—as if they were at prices no higher than 70 percent of crude oil prices (excluding severance taxes). Thus an existing (pre-1985) gas contract priced at the equivalent of distillate oil would be allowed to operate on its own terms, but would be treated *as if* it were priced at 70 percent of the crude oil price in affecting any other contract. No limits would be placed on *new* gas contracts signed after the date of decontrol. If a new contract were signed at 80 percent of the crude oil price, it would be allowed to trigger old contracts that refer to it. The old contract would, however, be treated as if it were priced at only 70 percent of the crude oil price in triggering other contracts.

The option has the advantage of limiting the effect of ill-advised old gas contracts but allowing new gas contracts (signed in a deregulated gas market) to "communicate" market information to older gas contracts. It relies on market forces to modify the oil parity contracts themselves.

Option 5: the Gramm proposal. Representative Philip Gramm (Democrat, Texas) has proposed a contract solution that places a ceiling on the operation of indefinite price escalators based on the average price of newly contracted gas in a given region (see H.R. 5866). Any contract signed before the date of enactment and containing an indefinite pricing provision would be subject to the ceiling. Already deregulated gas would be frozen at current contract prices until the gas cap sets a higher price. Renegotiated contracts and new contracts would not be subject to the cap. Unlike options 2 and 4, Gramm's proposal ties the cap to fluctuations in gas prices (both increases *and* decreases) rather than in oil prices.

The proposal is designed to use a market-based measure of the clearing price for natural gas, rather than a legislator's guess. The proposal assumes, however, that prices for new gas contracts will not be affected by continued price controls on some gas (Gramm's bill phases out all price controls by 1985). This problem is more severe under a longer, rather than a shorter, phase-out of price controls.

A major unanswered question is whether the implementation problems associated with measuring the regional price for new gas contracts will be manageable. Not all prices are comparable. A gas price may or may not include the costs of gathering lines, for example, or the costs of gas processing. Lags in data collection and the difficulties of defining appropriate regions could complicate the measurement problem. Although the Gramm approach is conceptually appealing because of its reliance on market signals, these implementation problems deserve careful scrutiny.

Conclusion

The president has decided that the administration's legislative agenda is too crowded to afford room for an administration bill on natural gas pricing in 1982. The debate over the workability of the NGPA, however, is likely to continue. In contrast to the debate in 1977–1978, this year's issues will focus on the validity of the market-ordering problems. Can partial decontrol be made to work, or will the unequal access to price-controlled gas foster regional bidding wars? Will the gas industry be able to renegotiate its own contracts to adjust to changing market conditions, or will the heavy hand of the federal government be required to "ease the transition"? Finally, and most significant, will the solutions to these market-ordering problems rely on market-like mechanisms, or will they impose more regulation to solve the transition problems? The lesson from the NGPA is that regulation fosters more regulation: the pitfall in the current debate is the temptation to use more regulation to solve the "transition" problems on the road to deregulation.

Notes

1. U.S. Department of Energy, Office of Policy, Planning, and Analysis, Division of Energy Deregulation, *A Study of Alternatives to the Natural Gas Policy Act of 1978* (November 1981) (hereafter cited as the DOE study).
    2. The DOE study assumed a world oil price of $35 per barrel (in 1980 dollars) in 1985 as a reference base. Sensitivity analysis was also performed on higher and lower world oil prices, ranging from $31.43 to $40.00 per barrel in 1985.
    3. Energy Information Administration (EIA/DOE), *The Current State of the Natural Gas Market: An Analysis of the Natural Gas Policy Act and Several Alternatives, Part 1* (December 1981), p. 51.
    4. American Gas Association, "Consumer Impact of Indefinite Price Escalator Clauses under Alternative Decontrol Plans" (November 6, 1981). The AGA's estimate may also include gas imports, which receive higher prices than domestic supplies.
    5. All prices are in mid-1980 dollars unless otherwise noted.
    6. American Gas Association, "The Spring 1982 A.G.A.–TERA Base Case," July 13, 1982.
    7. Because the NGPA specifies price ceilings that exclude severance and other

taxes, the gas prices used to calculate the ratio between oil and gas prices exclude taxes. The percentage price increases *include* taxes.

8. For details of the different treatment of interstate and intrastate gas, see the DOE study, p. 7.

9. Energy Information Administration, *Intrastate and Interstate Supply Markets under the Natural Gas Policy Act* (May 1981), pp. 11, 16, tables 3, 8. Prices are as reported in EIA's December 1978 survey.

10. EIA, *The Current State of the Natural Gas Market* (December 1981), pp. 65–66. Prices in this section are as reported to FERC for late 1981 and early 1982.

11. Summary of statement of Henry B. Taliaferro, Jr., representing GHK Companies and the Independent Gas Producers Committee before the Subcommittee on International Trade, Finance, and Security Economics, Joint Economic Committee, February 18, 1982, p. 5.

12. The American Gas Association has attempted to analyze bidding behavior in today's gas market (see AGA, "A Statistical Analysis of Bidding Trends for Decontrolled Natural Gas under the NGPA," March 19, 1982). Unfortunately the analysis is flawed by its failure to account for these complexities in measurement of the cushion.

13. Two of these pipelines have subsequently become large purchasers of 107 gas. By mid-1982, El Paso's 107 purchases rose to 8 percent and Transwestern's to 15 percent. Panhandle Eastern's and Northern Natural's purchases have remained low. See Energy Information Administration, *An Analysis of Post-NGPA Interstate Pipeline Wellhead Purchases* (August 1982).

14. AGA, "Consumer Impact," p. 3. More recent AGA forecasts project a loss of only 0.6 quadrillion Btu (roughly equivalent to 0.6 trillion cubic feet) in 1985. See AGA, "Spring TERA Case," p. 2.

15. Written statement of the American Gas Association before the Joint Economic Committee, February 18, 1981.

16. Merrill Lynch, *Natural Gas Monthly: Outlook for 1982* (January 1982), p. 6.

17. Energy Information Administration, *Natural Gas Producer/Purchaser Contracts and Their Potential Impacts on the Natural Gas Market*, June 1982.

18. AGA, "Consumer Impact," p. 3.

19. Interstate Natural Gas Association of America, *Analysis of Natural Gas Decontrol*, December 1981, p. 1.

20. EIA, *Intrastate and Interstate Supply Markets*, p. 19.

21. The assessments are entirely those of the authors and should not be construed as positions of the Department of Energy or the Interstate Natural Gas Association of America.

22. See "Supplemental Statement on Behalf of the Interstate Natural Gas Association of America before the Senate Committee on Energy and Natural Resources on Implementation of Title I of the Natural Gas Policy Act of 1978," December 7, 1981.

23. Federal Trade Commission, staff report of the Bureau of Economics, *The Economic Structure and Behavior in the Natural Gas Production Industry* (February 1979), p. 3.

The Intrastate Pipelines and the Natural Gas Policy Act

Robert C. Means

Intrastate pipelines face a doubtful future. The doubts have three sources. One is the transition to a national market in which price plays its customary role in allocating supply between customers and between regions. In that national market the marginal consumer in Detroit and New York will compete for natural gas with the marginal consumer served by the intrastate pipelines, and it is unclear how the latter will fare in this competition.

A second source of uncertainty concerns the intrastate pipelines' traditional sources of supply. Reserves are declining in both Texas and Louisiana, as they are in most of the traditional major producing areas, and most of the new frontier areas lie outside the major producing states. Even if intrastate pipelines are able to compete on an equal footing with interstate pipelines, they will be competing for a diminishing resource base unless they can gain effective access to natural gas outside their borders.

Both these sources of uncertainty would exist even under complete decontrol. To them the Natural Gas Policy Act of 1978 (NGPA)[1] adds the potential problem of bidding disparities arising from the differential effects of the act's price ceilings and deregulation schedule on interstate and intrastate pipelines. This last problem is the subject of this paper.

An Initial Overview

The Legal Basis of the Dual Market. The NGPA itself was in part a response to interregional bidding disparities. The source of the disparities was the existence of statutory limits on the price that interstate pipelines could bid for gas but not on the price bid by the intrastate pipelines with which they compet-

Edwin Malet of the Office of Regulatory Analysis and Jeffrey Price, formerly of that office, assisted in the preparation of this paper. The views expressed in the paper are the author's and are not necessarily those of the Federal Energy Regulatory Commission. Sections of this paper were written after the conference.

ed for supplies. The result was what came to be known as the dual market: a division of the field and burner-tip markets into largely separate interstate and intrastate segments. The NGPA fundamentally altered the nature of the bidding disparities between natural gas pipelines. The dual market, however, continues to exist.

The origin of the dual market lay in the historical development of producer price regulation under the Natural Gas Act (NGA).[2] In *Phillips Petroleum Co. v. Wisconsin,* the Supreme Court held that natural gas producers that sell gas to interstate pipelines are themselves "natural gas companies" within the meaning of the act.[3] Those producers thus became, in a formal legal sense, public utilities. The most important economic consequence of this definitional sleight-of-hand was to subject them to federal price regulation, and much of the subsequent history of that regulation can be explained by the tension between the traditional rate-making methods of public utility regulation and the economic reality of the natural gas producing industry.[4]

But the definition also meant that natural gas producers were subject to regulation of market entry and exit under section 7 of the NGA. A producer that sold natural gas in interstate commerce was required to obtain a certificate of public convenience and necessity from the Federal Power Commission (FPC) and could not then abandon the service without the commission's approval.

The practical implications of certification for natural gas producers were substantially different from the implications for conventional public utilities. What was certified was not a business enterprise but a resource base—for example, natural gas reserves underlying a certain acreage. Reserves covered by a certificate were commonly referred to as being dedicated to interstate commerce. Although a producer could not withdraw dedicated reserves from interstate commerce without the FPC's approval, he was generally under no legal obligation to add new reserves to offset the decline of those already dedicated. If he found it more profitable to sell new, undedicated reserves in intrastate commerce, or indeed more profitable to direct his labor and capital outside the natural gas industry, he was legally free to do so.[5]

Nevertheless, dedication did mean that there was a reasonably well-defined distinction between natural gas that was subject to federal wellhead price controls and natural gas that was not. The former was commonly called interstate gas because it was deemed to be sold in interstate commerce, although it might sometimes in fact be sold within the state of its production. The latter was commonly called intrastate gas.

The distinction between interstate and intrastate natural gas was closely linked to two other distinctions: that between interstate and intrastate pipelines and that between the interstate and intrastate burner-tip markets. The first of these also rested on section 7 of the NGA. An interstate pipeline is one that holds a certificate of public convenience and necessity, and a certificate is in turn generally required if the pipeline transports natural gas in interstate commerce.[6] An intrastate pipeline is in general one that does not transport gas in interstate commerce and therefore does not need a section 7 certificate.

Under the NGA, natural gas generally had to be dedicated to interstate commerce if it was sold to an interstate pipeline, and a pipeline generally had to obtain a section 7 certificate if it purchased or transported natural gas that was dedicated to interstate commerce. To avoid federal regulation, intrastate producers and intrastate pipelines thus had to form a closed system, dealing only with one another and not with interstate producers or pipelines.[7] A corollary of the intrastate system's closure was that interstate producers and pipelines also formed a closed system.

The isolation of the interstate and intrastate systems from each other made it possible to speak of separate interstate and intrastate natural gas burner-tip markets. The interstate market was served by natural gas that was subject to federal wellhead price regulation and was transported by pipelines likewise subject to federal regulation. The intrastate market was served by producers and pipelines that either were unregulated or were regulated only by the states. The separation of the burner-tip markets was less clear cut than the separation of the systems that served them, since some geographical areas and even some users were served by both interstate and intrastate pipelines. Nevertheless, the distinction was clear enough for most practical purposes.

The NGPA left federal pipeline regulation largely intact and thus preserved the NGA's implicit distinction between interstate and intrastate pipelines. The effect of the NGPA on the distinction between interstate and intrastate production was more complex. The NGPA abandoned the pretense that natural gas producers are public utilities. Although price controls are retained on nearly all natural gas until 1985, the price ceilings are for the most part determined by statutory formulas rather than by the cost-based methods of public utility price regulation,[8] and the Federal Energy Regulatory Commission (FERC) no longer issues certificates of public convenience and necessity for natural gas production. Producer certificates outstanding at the time of enactment of the NGPA remain in force, however, and the natural gas covered by those certificates continues to be dedicated to interstate commerce.[9] More important, the price and deregulation provisions of the NGPA depend in significant ways on the classification of the natural gas before the NGPA was enacted.

The NGPA establishes incentive prices for natural gas from post-NGPA wells (sections 102, 103, 107) and from wells with very low rates of production (section 108). Qualification for these sections depends only on the characteristics of the well and reservoir and not on whether the natural gas was previously dedicated to interstate commerce.[10] But the price ceilings for natural gas from pre-NGPA wells, often called old or flowing gas, were intended not so much to create incentives as to preserve the pre-NGPA status quo; for the categorization of flowing gas, the question of dedication to interstate commerce is crucial. Three sections govern such gas: sections 104, 105, and 106. If the gas was dedicated to interstate commerce, it is regulated by section 104 (existing and successor contracts) or section 106(a) (rollover contracts); if it was not so dedicated, it is regulated by section 105 (existing and successor contracts) or section 106(b) (rollover contracts).

For the rollover contracts governed by section 106, the price difference between interstate and intrastate gas is explicitly set out in the statute: a minimum price of 54 cents, adjusted for inflation since April 1977, if the gas was dedicated to interstate commerce; a minimum price of $1.00, similarly adjusted, if it was not so dedicated. The more important distinction, however, is between sections 104 and 105. The price ceiling is not expressly set out in those sections. Rather, the ceiling is based on the price paid for the natural gas on April 20, 1977, for interstate gas (section 104) or on November 8, 1978, for intrastate gas (section 105). The pre-NGPA prices to which these two sections refer were significantly different in the interstate and intrastate markets. Sections 104 and 105 carry that difference forward into the NGPA's price structure.

The more significant effect of interstate dedication, however, is on deregulation. Two important categories of natural gas are deregulated only if they were not dedicated to interstate commerce in 1977 or 1978:[11] natural gas from new onshore production wells, which is deregulated on January 1, 1985 (wells deeper than 5,000 feet), or July 1, 1987 (wells of 5,000 feet or less);[12] and natural gas that sells for more than $1.00 on December 31, 1984.[13]

The relative amount of natural gas still subject to price controls will therefore be substantially larger for interstate pipelines than for intrastate pipelines. The Energy Information Administration (EIA) of the Department of Energy estimates that approximately 58 percent of interstate natural gas that is price-controlled in 1984 will still be controlled after January 1, 1985, in comparison with approximately 36 percent for the intrastate market. In 1987, when shallow intrastate section 103 gas is deregulated, the percentage of price-controlled gas in the intrastate market will be reduced even further.[14]

In sum, under the NGPA the average price of price-controlled gas is now generally lower for interstate market pipelines than for intrastate ones. After 1985 the average *price* of price-controlled gas will continue to be lower in the interstate market, but the more important difference is that the relative *amount* of price-controlled gas will be much larger for the interstate market than for the intrastate market. These differences in the price and quantity of price-controlled gas are reflected in the relative size of the interstate and intrastate natural gas cushions, and it is the distribution of the cushion between the interstate and intrastate markets or pipelines that is the basis for the bidding disparities discussed in this paper.

The bidding disparities between interstate pipelines on the one hand and intrastate pipelines on the other are not the only disparities that exist in today's natural gas markets or will exist in the future. There also exist significant differences among interstate pipelines and among intrastate pipelines with respect to the price and volume of price-controlled gas. This paper's focus on the bidding disparities between interstate and intrastate pipelines is in part a necessary accommodation to the available information, much of which is based on a division of the national gas market into interstate and intrastate segments. But the interstate-intrastate bidding disparity also has a different character from the disparities within each group of pipelines. The latter disparities are at least in part the result of differences in past circumstances and management practices. The interstate-intrastate bidding disparity is primarily a product of the different regulatory history of the two groups of pipelines.

The Effect of the Dual Market. The legal distinctions between interstate and intrastate gas for price ceilings and decontrol affect the relative size of the natural gas cushion controlled by the interstate and intrastate pipelines. The cushion is the difference between the actual cost of price-controlled gas to pipelines and the amount they would pay for the same amount of gas under total decontrol. It measures, in other words, the amount of revenue lost by producers of price-controlled gas as a result of controls.

If all natural gas were subject to wellhead price controls, the producers' loss would be the consumers' gain: the economic benefit of the cushion would be enjoyed by consumers through lower prices. This benefit would have its costs: natural gas production would be smaller, and it would also generally be necessary to ration natural gas in some manner because the price would not be high enough to balance supply and demand. These costs are, of course, precisely the ones commonly associated with price controls that are not limited to the elimination of monopoly profits, and they were in fact the cost of price controls under the NGA.

The implications of the natural gas cushion under partial price controls are quite different.[15] If pipelines do not exercise collective self-restraint in the natural gas field market—if they bid competitively for new supplies, in other words—the immediate effect of the natural gas cushion will be a price for deregulated gas that is higher than it would be under complete decontrol. The reason is that pipelines commonly price gas to their customers at the pipelines' average cost. The marketability of natural gas thus does not depend on the price paid for any particular part of the supply. If the price for a portion of the supply is held down by price controls, pipelines can pay more for the remaining supply without being unable to market the gas. And if the field-market price of deregulated gas is in fact determined by competitive pipeline bidding, the pipelines' ability to resell the gas is the only constraint on that price.[16]

If the natural gas cushion were distributed evenly among pipelines—I pass over for now just what "evenly" means in this context—the consequences of partial decontrol would generally differ from those of total decontrol as follows. The price of regulated gas would be lower and the producers of that gas poorer, and the price of deregulated gas would be higher and its producers richer, than under complete decontrol. In addition, if the supply of deregulated gas is more responsive to price than the supply of regulated gas, the total supply would be larger than under complete decontrol: the supply lost as a result of the lower price for regulated gas would be smaller than the supply gained as a result of the higher price (than under complete decontrol) for the unregulated gas.

On the consumers' side, all would confront a market-clearing price under partial decontrol, just as they would under complete decontrol; this result follows from the assumption that *each* pipeline continues to bid for additional natural gas so long as it can resell the gas. If the supply is larger under partial than under complete decontrol, that market-clearing price presumably would be somewhat lower than the complete-decontrol price. Interregional price differences and distribution of supply, however, should be substantially the same as under complete decontrol.

In essence, with competitive field-market bidding and an evenly distributed cushion, partial decontrol can best be viewed as a tax imposed on producers of regulated gas to support a subsidy for unregulated gas. Natural gas consumers do not benefit directly from the control but benefit indirectly if the subsidy results in increased production. Whether society as a whole gains from the partial controls depends on whether the subsidy is efficient, in the sense of offsetting other market imperfections.

For reasons discussed in the preceding section, however, the natural gas cushion will not be evenly distributed between interstate and intrastate pipelines. The relative distribution of the cushion is discussed in the last part of this paper, "The Dimensions of the Problem." Here it is sufficient to note that interstate pipelines now generally have a substantially larger cushion than intrastate ones, and the interstates' advantage will increase significantly on January 1, 1985. This uneven distribution of the cushion will not necessarily lead to different results as measured by national averages. It will, however, lead to different results from the standpoint of particular regions and pipeline service areas.

The price of unregulated gas will presumably be relatively uniform nationally. The average cost of natural gas to each pipeline, however—and thus the purchased-gas cost to its customers—will vary inversely with the size of the pipeline's cushion. This price differential will in turn produce a supply shift, as marginal customers on shallow-cushion pipelines turn to alternative fuels or to other pipelines while deep-cushion pipelines are able to expand their markets by competing with lower-valued fuels or by acquiring customers formerly served by pipelines with smaller cushions.[17] These supply shifts may reduce the price differential, as the benefit of the price-controlled gas of the shallow-cushion pipelines is spread over a smaller volume while that of the deep-cushion pipelines is spread over a larger one.[18] In addition, as the supply shifts, deep-cushion pipelines will serve additional customers to whom natural gas is less valuable, and shallow-cushion pipelines will be left with higher-valued uses at the margin. At some point the differences in the marginal values of natural gas in the burner-tip markets should balance not only the genuine economic differences in transmission costs but also the artificial cost differences caused by the uneven distribution of the cushion. It is precisely in that balance, however, that the direct efficiency costs of the uneven distribution appear, as lower-valued uses served by deep-cushion pipelines can obtain gas while higher-valued uses served by shallow-cushion pipelines cannot.

A Closer Look at Assumptions and Concepts

Intrastate Markets and Intrastate Pipelines. This paper is concerned with the effect of the interstate pipelines' cushion on the price and allocation of natural gas. For present purposes, therefore, the intrastate market can be defined as the market that does not benefit directly or indirectly from that cushion. Direct industrial sales by interstate pipelines are therefore excluded from the intrastate market even though such sales are not subject to federal price regulation.[19] Also excluded are customers served indirectly by interstate pipelines through local distribution companies and Hinshaw pipelines.[20] The intrastate market then consists of three kinds of end user: those served by ordinary intrastate pipelines, either directly or indirectly through distribution companies; those served by Hinshaw pipelines that make only limited or irregular purchases of gas from interstate pipelines;[21] and those that purchase gas directly from producers.[22]

The bidding disparity problem discussed in this paper is not equally relevant to all parts of this intrastate market. The problem appears to be most acute in the major producing states of Louisiana and Texas. Within those states, it is a problem of the intrastate pipelines; it is not, at least in the form discussed above, a problem of the end users that purchase natural gas directly from producers.

State markets. The intrastate market is in fact a collection of individual state markets. For interstate pipelines, state boundaries are essentially irrelevant. An interstate pipeline may confine its operations to certain field or burner-tip markets for economic reasons or because it cannot obtain the regulatory approval needed for expansion; it will seldom if ever do so simply because expansion would carry its operation into new states. For intrastate pipelines, however, state boundaries are of critical importance. An intrastate pipeline in one state cannot compete directly for supplies in another state, and only an intrastate pipeline qualifying for the Hinshaw exemption can compete even indirectly for out-of-state supplies. In the burner-tip market, intrastate pipelines can sell outside their own state only within the limits of sections 311(b) and 312 of the NGPA.[23]

The separate state intrastate markets share a common history of freedom from federal regulation under the Natural Gas Act, and that common history has significant implications for their regulation under the NGPA. Each market has also been affected, however, by local supply and demand and in some cases also by state regulation.

An intrastate market requires local gas production and local buyers. Thirty-one states had some natural gas production during 1980,[24] and according to a survey of 1978 sales, some production was purchased by intrastate pipelines or directly by end users in at least twenty-four of them.[25] Intrastate sales volume was small in a number of cases, less than 20 billion cubic feet in nearly half the states. In contrast, 1978 intrastate sales in Texas were reported at 3.7 trillion cubic feet. Nevertheless, there was in some sense an intrastate market in two dozen states at the time the NGPA was enacted. The number may be smaller today; in particular, the interstate pipelines' improved bidding position may have made direct producer sales relatively less attractive.[26] The intrastate market still extends, however, well beyond the major producing states with which it is commonly associated.

Although no two of the intrastate markets are identical, it appears that they can be divided into three broad groups. One consists of the Texas and Louisiana markets. Texas and Louisiana are the principal gas-producing states; they are also the largest exporters of gas to other states and the first and third largest consumers of gas.[27] In both states, gas consumption is dominated by a large industrial and electric utility demand met mainly by intrastate pipelines and direct producer sales.[28] Before enactment of the NGPA, competition to serve this market drove the unregulated wellhead price for intrastate gas well above the regulated interstate level. For gas from old wells, the NGPA then froze this price difference between interstate and intrastate gas, leaving most Louisiana and Texas intrastate pipelines with current average gas costs higher than those of the major interstate pipelines.[29] That price difference is likely to become larger in 1985 as a result of the uneven effect of partial deregulation on interstate and intrastate pipelines.

A second group of intrastate markets consists of the three next largest net gas exporters: New Mexico, Oklahoma, and Kansas.[30] Demand by industries and electric utilities for natural gas is less significant in these states than in Texas and Louisiana, and competition among intrastate pipelines to acquire gas appears to have been less intense.[31] Intrastate prices in these states were generally lower at least than those in the Texas intrastate market on the eve of enactment of the NGPA,[32] and the legislatures in these states reacted to the NGPA by restricting the movement of prices under old intrastate contracts to NGPA levels.[33] This state price control legislation has been challenged in the courts.[34] The price controls expire in New Mexico on July 1, 1983, and in Kansas and Oklahoma on December 31, 1984.

As a result of different pre-NGPA market conditions and current state legislation, the intrastate pipelines in this second group of states appear now to stand in a substantially different position from those of Texas and Louisiana. Expiration of the state price control legislation will presumably narrow the difference, but even then the lower pre-NGPA intrastate prices in the second group of states may continue to give their intrastate pipelines a relatively larger cushion than those of Texas and Louisiana.

The third group consists of the remaining states in which producers sell gas to intrastate pipelines or directly to end users. These states are extremely diverse. In 1978 California and Colorado had intrastate markets larger than that of New Mexico; others had intrastate markets only a tenth that size.[35] They are lumped together here because too little is know about them to permit significant distinctions to be drawn. Intrastate sales in some of the states were probably too small to support an autonomous intrastate market. Nevertheless, in 1978 intrastate pipelines operated in all the states except perhaps California.[36] In the absence of state legislation, the bidding position of these pipelines will presumably be generally similar to that of Texas and Louisiana, at least after 1985.

Direct sales and pipeline sales. Not all natural gas in the intrastate market is bought and sold by pipelines; substantial volumes are sold directly by producers to end users.[37] In 1978 direct producers' sales accounted for nearly a third of the Texas intrastate market and half of the Louisiana intrastate market. Many of the direct sales in these two states are covered by contracts that antedate the large price increases of the 1970s,[38] and the price is generally low, substantially lower on average than the price paid by Texas and Louisiana intrastate pipelines.[39] The end users receiving gas under these contracts are thus the beneficiaries of a cushion that is relatively larger than those of most intrastate pipelines and indeed larger than the cushion controlled by the interstate pipelines. The policy significance of this end-user cushion is, however, quite different from that of a pipeline cushion.

The difference is not principally that only a single firm benefits from a direct sale by a producer. Some intrastate pipelines serve relatively few customers; the number of persons having an economic interest in a single large industrial purchaser may be very large. The principal difference lies, rather, in the relationship between the low-cost gas and current field bidding. For pipelines this relationship rests on the general practice of selling gas at average cost.[40] Given this practice, low-cost gas in the hands of a pipeline increases the price that it can rationally pay for new supplies. Direct purchasers, however, are generally free to maximize profits and in theory therefore should be concerned only with the marginal cost of gas, not with its average cost. In practice there may be circumstances in which the existence of a low-cost contract will affect a firm's willingness to pay for costly new supplies.[41] The link between the cushion and current bidding behavior is, however, much more speculative and indirect for direct purchasers than for pipelines.

Unlike the distinction between separate state intrastate markets, that between pipeline and direct producer sales within the intrastate market may be a temporary one. More than half the contracts governing direct producer sales are expected to expire by 1985.[42] The direct purchasers will then have two alternatives. They can renew their producer contracts, or they can turn to intrastate or interstate pipelines for their supply. The first alternative could be very costly, since the purchaser may no longer have any low-cost gas with which to average down the high price of new supplies. The second alternative would result in a very large addition to the demand on the intrastate or interstate pipelines concerned.

Too little is now known about direct sales to do more than speculate about the possibilities, however. Perhaps the only point that can be made with certainty is that such sales have characteristics very different from those made to the pipeline segment of the intrastate market and that the two segments must be treated separately in analyzing the bidding disparity issue.

The Competitive Bidding Assumption. The concept of bidding disadvantage or disparity is meaningful only if pipelines in fact acquire gas by competitive bidding. Pipelines of course bid for gas in the sense that they acquire new supplies by offering price and other terms sufficiently favorable to producers. The assumption that pipelines acquire gas through *competitive* bidding goes beyond this obvious point, however. It implies that a pipeline seeks to obtain additional gas by outbidding other pipelines. And since the other pipelines are seeking to do the same, competitive bidding will raise the price of gas to the point where some external constraint is reached. For regulated gas, that constraint will generally be the price ceiling. For unregulated gas, it will be the pipeline's ability to resell the gas.

The competitive bidding assumption thus implies that pipelines will bid the price of unregulated gas to the point where supply and demand are in balance, both nationally and on each pipeline. This assumption is critically important to the issue of bidding disparity because it establishes the causal link between the distribution of the natural gas cushion and the distribution of the natural gas supply. Its significance is perhaps best seen by examining the implications of the counterassumption—that pipelines do *not* bid the field price to a market-clearing level.[43]

The counterassumption might take various forms. It might be assumed, for example, that under partial decontrol pipelines will bid the price of unregulated gas above the complete-decontrol price but not necessarily high enough to clear the market. A counterassumption of this kind, based on the fact that pipelines purchase gas under long-term contracts, is discussed below in the section "The Bidding Advantage." For the moment, however, I will take the counterassumption in its simplest form: that pipelines bid the price of unregulated gas up only to the complete-decontrol level. On that assumption, consumers receive the full direct benefit of the natural gas cushion in the form of lower burner-tip prices, and average purchased-gas costs vary from pipeline to pipeline in inverse relation to the size of their respective cushions.[44] Rationing of some form is necessary on some or all pipelines, since by hypothesis the field-market price has not been bid up to the market-clearing level. As a result, the distribution of natural gas supply between pipeline systems is indeterminate: at least some pipelines could absorb additional supply at the prevailing price, but they cannot attract additional supply by bidding a higher price without violating the assumption that they do not bid above the complete-decontrol level.

On this counterassumption, intrastate pipelines and other pipelines with a relatively small cushion are at a disadvantage in the sense that purchased-gas costs are higher for their customers than for customers of better-endowed pipelines. Even for customers of intrastate pipelines, however, the burner-tip price is lower than it would be under complete decontrol, and the intrastate pipelines do not necessarily lose supply to their interstate competitors.

The counterassumption just described is clearly invalid for the current natural gas market.[45] It is useful principally as a limit: it marks the point where a deep cushion ceases to confer any advantage in acquiring new supplies of natural gas. Under more realistic counterassumptions, such as the one discussed in "The Bidding Advantage," a deep cushion does confer an advantage in this respect, although the advantage may be smaller or less clearly defined than it is under the simple competitive bidding assumption.

The Concept of Bidding Disparity. What does it mean to say that the intrastate pipelines are (or are not) at a disadvantage in bidding for natural gas? The statement evidently implies a comparison, but a comparison with what? One possible comparison is historical. Intrastate pipelines undoubtedly are worse off now than they were in the early 1970s, when federal price controls effectively foreclosed the onshore field market to interstate pipelines. Intrastate pipelines then had to compete only with one another; not surprisingly, they generally found it easy to acquire ample reserves at a price that allowed them to resell the gas to industrial customers with a comfortable markup. In comparison with the pre-NGPA period, any legal regime that permits interstate pipelines to bid effectively for new natural gas supplies is bound to make life more difficult for intrastate pipelines, forcing them to pay a higher price for smaller quantities of gas than if they still had the field market to themselves. In this sense at least, intrastate pipelines are now almost certainly at a bidding disadvantage, but if this is the only sense in which they are disadvantaged, their problems have little or no policy significance; the disadvantage is simply the inevitable consequence of the creation of a single national market for natural gas. In such a national market, intrastate pipelines—and therefore their customers as well—can be expected to fare less well than they did in the dual market that existed before 1978. For policy purposes, however, the relevant question is, Do the NGPA's adverse consequences go beyond those that must accompany the transition to a rational and nationwide market?

This question suggests that the appropriate point of comparison is a national market free from any institutional distortions. The reference point used in this paper is the price and distribution of natural gas under complete decontrol. There is bidding parity if interstate and intrastate pipelines are able to acquire the same relative proportions of the total supply of natural gas that they would have under complete decontrol.[46]

If interstate pipelines had a cushion and intrastates had none, it would be easy to say that the intrastate pipelines were at a bidding disadvantage—that they would be worse off than they would be under complete decontrol. Intrastate pipelines have their own cushion, however, and it is therefore necessary to compare the estimated distribution of the cushion with a distribution that would result in bidding parity between the two pipeline systems.

Two measures of bidding parity appear to be potentially useful. The first I will call *pro rata parity*. Pro rata parity is measured by dividing the interstate and intrastate pipelines' respective cushions by their respective volumes of natural gas under complete decontrol. For example, suppose that interstate pipelines have a cushion of $10 billion and would have 12 trillion cubic feet (Tcf) of natural gas under complete decontrol and that intrastate pipelines have a $5 billion cushion and a complete-decontrol volume of 6 Tcf. The interstate pipelines have a larger cushion, but in relation to the complete-decontrol volume of the two systems, the cushions are identical: $0.83 per thousand cubic feet (Mcf). By the pro rata standard, there is no bidding disparity between the interstate and intrastate pipeline systems.

Pro rata parity measures pipelines' ability to share in the supply of deregulated gas in proportion to their share of the total natural gas supply under complete decontrol. To continue with the example, suppose that the total supply and, therefore, the market-clearing price are the same under partial and total decontrol: 18 Tcf and $4 per Mcf. Suppose further that under partial decontrol interstate pipelines would control 6 Tcf of price-controlled gas and intrastate pipelines would control 3 Tcf. On those assumptions, the price of decontrolled gas under partial decontrol would be $5.67, and the average price for each system ($4) and the distribution of supply between the two systems (12 Tcf and 6 Tcf respectively) would be the same as under complete decontrol.[47]

In this hypothetical illustration, pro rata parity does in fact place the two systems in the same relative position as complete decontrol. It does so, however, only because each system has the same relative amount of price-controlled gas: one half of its complete-decontrol volume. The addition of proportionate amounts of decontrolled gas to that price-controlled gas therefore results in the same distribution of total supply as under complete decontrol.

After 1985, however, the relative amount of natural gas still subject to price control will be substantially larger for interstate than for intrastate pipelines. If intrastate pipelines then share in the supply of decontrolled gas only in proportion to their complete-decontrol volume, they will be worse off than they would be under complete decontrol. To have the same proportion of total supply as under complete decontrol, they must be able to buy a disproportionately large share of the decontrolled gas.

To state the point more generally, a pipeline's bidding position is determined not only by the *dollar value* of its cushion but also by its *volume* of price-controlled gas. The former determines the amount that the pipeline has to spend in bidding for decontrolled gas; the latter determines the amount of decontrolled gas on which the cushion must be spent. The latter variable could be ignored if all pipelines had the same relative amount of price-controlled gas, but they do not. Another measure of bidding disparity therefore is required. The one used here I will call *need parity*. It is computed by dividing the pipeline's cushion by the difference between its complete-decontrol volume and its volume of price-controlled gas. Need parity thus measures a pipeline's ability to pay a price in excess of the complete-decontrol price for the amount of decontrolled gas needed to achieve its complete-decontrol volume; that needed amount is in turn equal to the difference between the complete-decontrol volume and the pipeline's volume of price-controlled gas.

"Decontrolled gas" in this definition includes all gas other than price-controlled gas: not only decontrolled gas from conventional domestic sources but also imported gas and synthetic gas. Intrastate pipelines receive little or no gas from these sources, but they constitute a significant part of the interstate supply. The definition of need parity implicitly assumes that the cost of gas from these sources is the same as the average field price of domestic unregulated gas. If these other sources are cheaper than domestic decontrolled gas, need parity understates interstate pipelines' bidding advantage. If they are more expensive, need parity overstates their advantage.

To illustrate need parity, it will be necessary to modify somewhat the example used earlier. As in that example, suppose that the interstate cushion is $10 billion but that interstate pipelines control 7 Tcf of price-controlled gas, in relation to the 5 Tcf of gas needed to give them their complete-decontrol volume, their cushion thus amounts to $2 per Mcf. Interstate pipelines can, in other words, pay a price for deregulated gas that is $2 above the $4 market-clearing price.

Suppose then that intrastate pipelines control only 2 Tcf of price-controlled gas and therefore need 4 Tcf of deregulated gas. Given the interstate pipelines' cushion and the distribution of the supply of price-controlled gas, intrastate pipelines would need a cushion of $8 billion to achieve need parity. With that cushion they can match an interstate bid of $6 per Mcf for decontrolled gas, and both systems will be left with the same supply and average price as under complete decontrol.

These arithmetical exercises obviously tell us nothing about the real world. The point they do make is that discussion of the intrastate pipelines' putative bidding disadvantage requires at the outset some careful definition of terms. The statement that interstate pipelines have a larger cushion than intrastate pipelines is almost certainly true but also almost valueless as evidence that the NGPA distorts the regional allocation of natural gas. Distortion implies an undistorted reference case. Presumably that point of reference should be the distribution of natural gas under complete decontrol.[48] The question of the intrastate pipelines' bidding disadvantage is thus ultimately a question of whether they can acquire the same share of the total natural gas supply as under complete decontrol.

Definitions and concepts are useful in posing this question. To answer it, however, or even to put some outer limits on the answer, it is necessary to turn from hypothetical illustrations to estimates and projections that may bear some relation to the real world. That is the purpose of the next part of this paper.

The Dimensions of the Problem

The proposition that the NGPA disadvantages intrastate pipelines in bidding for natural gas rests on two assumptions. The first is that natural gas supplies are allocated among would-be purchasers through competitive bidding; the second is that the distribution of the natural gas cushion places the intrastate pipelines at a bidding disadvantage. In theory, either assumption might simply be wrong: natural gas supplies might be allocated on some principle wholly unrelated to competitive bidding; if competitive bidding did play a part, intrastate pipelines might be at no disadvantage at all in the bidding process.

It seems reasonably clear, however, that the assumptions are not wrong in this sense. Competitive bidding does have something to do with the allocation of natural gas, and in this bidding intrastate pipelines are generally at some disadvantage. The relevant questions therefore are not ones of validity or invalidity but rather ones of degree. Competitive bidding does play a role in allocating natural gas supplies, but how much of a role? Intrastate pipelines are at a disadvantage, but how serious is it? It is with these two questions that the present section is concerned.

Pipeline Bidding Behavior. Earlier in this paper, two alternative assumptions concerning pipeline bidding were set out. One was the assumption on which the initial overview of the bidding disparity problem was based: that pipelines will bid up the price of deregulated gas until the average price of all gas balances supply and demand on each pipeline system. The other was the counterassumption that pipelines will not bid deregulated price above the complete-decontrol level.

The counterassumption clearly is not valid for the current market. Pipelines have not stopped bidding at the complete-decontrol level; nearly every major pipeline has purchased gas at a price that can be sustained only if averaged with less expensive regulated gas.[49] Part at least of the benefit of the natural gas cushion is going to producers of deregulated and tight sands gas. More to the point for this paper, purchases of this gas require a cushion of some minimum size.

Similarly, though less clearly, the competitive bidding assumption also may not be fully descriptive of the current market. Several recent studies suggest that current pipeline bidding may be influenced less by whether pipelines can afford to buy than by whether they need the gas to serve existing demand.[50] These studies raise three general questions for the bidding disparity issue. The first concerns their validity for the period from which their data were drawn, a period that runs generally from 1979 to early 1982. The second concerns their implications for the bidding disparity issue for that same period. Finally, the third question concerns the applicability of the studies to the current and future natural gas markets.

The studies examine the relationship between (1) pipelines' volumes of old gas and (2) their purchases of deregulated gas. All the studies show that the relationship is if anything a negative one: pipelines with large volumes of old gas are less likely than the average pipeline to be aggressive purchasers of deregulated gas. The apparent negative correlation is not interpreted as a perverse relationship between ability to pay and bidding behavior: pipelines do not bid aggressively for deregulated gas simply because they *cannot* afford to do so. Rather, the negative correlation appears to be simply a byproduct of the positive one between purchases of deregulated gas and pipelines' need for gas as measured by their reserve positions. Pipelines with relatively small volumes of old gas tend also to be relatively short of reserves in general, and their generally aggressive bidding for deregulated gas is apparently an attempt to remedy that reserve deficiency.

The studies provide useful insights into pipeline bidding behavior, but they do not directly address the competitive bidding assumption. The relationship implied by that assumption differs significantly from the one analyzed in the studies. The competitive bidding assumption implies that (1) the size of pipelines' *natural gas cushions* determines (2) the amount of natural gas that they acquire *relative to the amount that they would acquire under complete decontrol*. A pipeline's cushion is not the same as its volume of old gas. Section 103 or even section 102 gas may contribute to the cushion; more important, the size of a pipeline's cushion depends on the average price, as well as the volume, of its regulated gas. Nevertheless, the volume of old gas may serve as a reasonable surrogate for the natural gas cushion; there is in any event no reason to think that use of the old gas volumes biases the studies' results.[51]

To use purchases of deregulated gas as a measure of the current use of the cushion is more problematic. Deregulated gas purchases would constitute a reasonable measure for this purpose if (1) deregulated gas were the only gas currently selling above the complete-decontrol price and (2) the starting point for additional purchases were the distribution of natural gas that would prevail under complete decontrol. Tight sands gas and imports are also priced above the complete-decontrol price, however, and therefore absorb a part of the cushion of the pipelines that purchase them. More important, the existing distribution of natural gas among pipelines probably is quite different from the complete-decontrol distribution. Price controls permitted some pipelines to compete for industrial markets that they would be forced to give up under complete decontrol; on the other side, supply limitations caused by price controls prevented other pipelines from fully exploiting their marketing opportunities.

Even with no cushion or an evenly distributed one, therefore, higher prices and adequate supplies would lead to a redistribution of natural gas supplies among pipeline systems. The redistribution of natural gas supply under the NGPA thus can be viewed analytically as consisting of two distinct processes. One is the gradual elimination of the allocational distortions caused by the era of binding price controls. The other is the emergence of new allocational distortions as a result of the uneven distribution of the natural gas cushion.

For any pipeline these two processes may either reinforce or offset each other. A pipeline's ability to bid for uncommitted gas depends not only on the relative size of its cushion but also on whether past price regulation has left it overcommitted in markets that it cannot expect to maintain over the long term or with market opportunities that it could not previously exploit for lack of supplies. A pipeline with a very large cushion may be forced to use that cushion largely to retain a low-valued industrial load that it would lose under complete decontrol;[52] another pipeline with a smaller cushion may be able to bid aggressively for deregulated gas because it has greater marketing opportunities. In effect, the first pipeline is using its cushion to "bid" to retain the supplies needed to serve its existing low-valued industrial market, while the latter pipeline's bidding for new supplies is not supported by its (small) cushion so much as by its marketing opportunities.

To separate the effect of higher prices and adequate supplies from that of the uneven distribution of the cushion would require more information about pipeline burner-tip markets than is now available. Studies of the kind cited above therefore cannot be conclusive. The studies do, however, raise legitimate questions about the accuracy of the simple competitive-bidding assumption. The implications of those questions depend, however, on the details of the assumption that is made in its place. As indicated earlier, the simple counterassumption—that pipelines do not bid the price of deregulated gas above the complete-decontrol level—appears to be clearly wrong. A more plausible alternative might be called the intertemporal bidding assumption.[53] This assumption rests on the fact that both the natural gas cushion and purchases of more costly gas are embodied in long-term contracts. As a result the current use of the cushion determines not only how much gas a pipeline acquires today but also how much gas it will be able to bid for in the future. A pipeline that fully commits its cushion to the purchase of deregulated gas in one period may be unable to purchase any additional deregulated gas in a subsequent period and may indeed have difficulty in marketing the deregulated gas that it acquired earlier, as its supply of inexpensive old gas is gradually exhausted.[54]

Moreover, at least in theory a pipeline can husband its cushion for the future in another and more direct way. By reducing current takes from its low-cost sources of supply, it may increase the volume of gas available from those sources in the future. As a result not just the amount of *uncommitted* cushion but the *total* cushion controlled by the pipeline in the future may be increased.

Under the intertemporal bidding assumption, a pipeline therefore may have three general bidding strategies available to it:

Full commitment strategy: set current takes from low-cost sources at a high level[55] and use the cushion now to purchase additional supplies until the pipeline's average price balances supply and demand on its system.

Deferred commitment strategy: set takes from low-cost sources at a high level but stop current purchases of new supplies before the market-clearing level is reached.

Deferred production strategy: reduce current takes from low-cost sources.

The intertemporal bidding assumption changes the form of the bidding disparity problem but not its fundamental nature. Some pipelines may lack the minimum relative cushion needed to purchase any deregulated or tight sands gas; indeed some pipelines with very small cushions may be forced to release reserves of high-cost gas already committed to them. For other pipelines there is a choice of when the cushion is to be used.[56] A deep-cushion pipeline that does not use its cushion in current bidding does not thereby lose its advantage, however; it preserves its advantage for use in the future.[57] Similarly, a shallow-cushion pipeline that stays in the market today by using its cushion fully faces the prospect of being unable to purchase any gas above its market-clearing price in the future.

Whatever the studies' validity for natural gas markets in the immediate past, their applicability to the current and future natural gas markets is more doubtful. The data for the studies were drawn from a period in which the average price on all major pipeline systems was below the market-clearing price and deregulated gas constituted 3 percent or less of the total supply. The average price may now have reached the market-clearing price on some pipeline systems and be approaching it on others. The relative importance of deregulated gas has increased only slowly, but on January 1, 1985, it will increase suddenly to more than half the total supply and nearly all the uncommitted supply for which pipelines can bid. Both the present change in the balance between supply and demand and the prospective large change in the relative importance of deregulated gas are likely to limit pipelines' ability to pursue the deferred commitment and deferred production strategies. As a result, their purchases of deregulated gas are likely to be much more closely determined by their current ability to pay.

That ability in turn depends on the relationship of a pipeline's average gas costs to the market-clearing price level. In a simpler world, the concept and implications of a market-clearing price would be clear. At some price the amount of natural gas offered by a pipeline would just equal the amount that its customers were willing to buy. A pipeline for which average gas costs had reached that level clearly could not be pursuing a delayed commitment strategy: it might be deferring production from its low-cost reserves, but to the extent that those reserves were being produced, they would by definition be fully committed to offsetting the cost of its higher-priced gas.

In practice the concept of a market-clearing price and therefore its implications are considerably more ambiguous.[58] To begin with, the amount of natural gas that can be sold depends not only on a pipeline's average gas costs but also on the aggregate amount that burner-tip customers pay for transmission and distribution and on how that aggregate amount is distributed among customers through rate design. For a given level of average gas costs, sales can be increased either by reducing the average transmission and distribution margins or by shifting costs from more price-elastic to less price-elastic customers.[59] But even if transmission and distribution margins and rate design are considered fixed, there is still no well-defined market-clearing price. A market-clearing price is one at which demand and supply are equal. For natural gas, demand is reasonably unambiguous in that hypothetical equation; supply is not.[60]

Consider a hypothetical pipeline, Transregional Gas Pipeline (TGP). Wells now attached to TGP are capable of delivering 1.0 billion cubic feet (Bcf) of natural gas daily under normal conditions, and TGP's transmission lines from the natural gas field to storage fields located near its burner-tip markets are also capable of efficiently handling 1.0 Bcf daily. (Transmission lines from the storage fields to burner-tip markets have a larger capacity in order to handle peak demand.) TGP's aggregate take-or-pay obligation amounts to 0.75 Bcf daily, and it is impractical for TGP to reduce its daily purchases below 0.60 Bcf.[61]

On these facts, there are several possible candidates for the answer to the question, What is the TGP supply that should be compared with demand at a given price?

1. *Maximum current supply*, or 1.0 Bcf daily.
2. *Target supply:* TGP's management is likely to want some margin between current demand and maximum deliverability and also between current demand and the take-or-pay level. It may therefore regard its current supply as in balance with an average daily demand of, say, 0.85 Bcf.
3. *Take-or-pay level*, or 0.75 Bcf.
4. *Minimum practical purchase level*, or 0.60 Bcf.

The answer to the question depends on the purpose for which it is asked. The present purpose concerns pipelines' freedom to pursue alternative strategies in bidding for new supplies or in retaining the supplies already committed to them. A pipeline possesses some freedom of action as long as demand is below the level of maximum supply and above the minimum purchase level, but its freedom becomes increasingly limited as demand approaches either of these end points. At the upper end a pipeline has little choice but to bid aggressively for new supplies. At the lower end it becomes increasingly difficult for a pipeline to attach new reserves, and its cushion comes to be increasingly committed to holding down its average gas cost so that demand will not fall below the take-or-pay level or, beyond that, the minimum purchase level.

It is the latter situation that now appears generally to prevail. In the latter months of 1982 it is probable that demand on only a very few pipelines has fallen to the minimum practical purchase level. The great majority of pipelines thus retain some freedom to attach additional reserves or absorb further price increases, and in this sense one might say that supply and demand are not yet in balance. Demand on a majority of the major pipelines, however, appears to be at the take-or-pay level or below,[62] and probably for most of the other major pipelines it is below the target supply level. For the present at least, pipelines' ability to defer commitment of their cushion is becoming increasingly circumscribed.

The present commitment of the pipelines' cushion thus appears to be largely to avoid further load loss in the face of recession and declining oil prices. It is therefore reflected to only a very limited extent in purchases of deregulated gas.

Pipelines' freedom to bid or not to bid for deregulated gas also depends on the relative importance of that category of gas. Today it still constitutes only a very small part of the total supply. It forms a larger proportion of the supply available for purchase, but the greater part even of the uncommitted gas now coming on the market is still subject to statutory price controls. As recently as 1981, therefore, a number of interstate pipelines were able to avoid purchasing virtually any deregulated gas, and a number of others evidently were able to confine their deregulated purchases to noncompetitive markets.[63]

Already by mid-1982 these market conditions had changed substantially. Only a single major interstate pipeline was still reporting no section 107 purchases,[64] and the number of pipelines paying relatively low prices had also declined.[65] These changes presumably reflect the moderate increase in section 107 volumes over the year. On January 1, 1985, deregulated gas will increase to perhaps nearly 60 percent of the total supply. Nearly all uncommitted gas will then be deregulated; the category of uncommitted but statutorily price-controlled natural gas will virtually disappear.[66] Under those circumstances no pipeline will be able to ignore the market in deregulated gas, nor are pipelines likely to be able to confine their purchases to situations where they enjoy an unusually strong bargaining advantage.[67]

This does not mean that all pipelines will be competing for every uncommitted packet of gas in 1985. The location of pipelines in relation to the natural gas field largely determines the set of competing pipelines over the short term. Over the longer term pipeline companies may extend their lines into promising new regions, but even this potential competition does not ordinarily encompass all major pipelines. Competition for deregulated gas will effectively involve a significantly larger number of pipelines in 1985 than it does today, however, and in 1985 the competitors will include the deep-cushion pipelines that have until recently been largely able to avoid the competition.

To stop competitive bidding short of the market-clearing price in 1985, therefore, will require collective self-restraint under conditions that are not very favorable to its exercise. The self-restraint will, moreover, have to consist of something considerably more precise than the mere avoidance of clearly outrageous prices. With nearly 60 percent of the total supply free from price controls, the deregulated price needed in 1985 to produce a market-clearing average field price should not appear unreasonable in comparison with those then being determined by oil-based escalator clauses, and under virtually any conceivable market conditions it will be substantially lower than the prices now being paid for deregulated gas.

Pipelines cannot explicitly agree to hold down the field price of natural gas without violating the antitrust laws. The question, then, is whether they can be expected to exercise sufficient collective self-restraint in the absence of such an agreement. They are not doing so today: if the prices being paid for deregulated natural gas are insufficient to clear today's market for all pipelines, it is only because the amount of deregulated gas is not very large. The conditions for self-restraint will be less favorable in 1985.

The Bidding Advantage. Concepts may be tailored to fit either the data at hand or the policy questions posed. The concepts of pro rata parity and need parity are of the latter sort. They were developed in this paper to define more precisely the issues of interpipeline bidding raised by the NGPA. That genesis

TABLE 1
AVERAGE PURCHASE COST OF GAS: COMPARISON OF INTRASTATE AND
INTERSTATE PIPELINES, TEXAS, 1975–1981
(dollars)

	1975	1976	1977	1978	1979	1980	1981
Intrastate pipelines[a]							
Amoco	0.94	1.44	1.75	1.91	2.03	2.31	2.74
Valero Transmission	1.48	1.81	1.97	2.09	2.22	2.54	3.23
Lone Star Gas	n.a.	n.a.	n.a.	n.a.	1.82	2.04	2.75
Pioneer Natural Gas	0.71	0.95	1.16	1.38	1.73	2.01	2.40
United Texas Transmission	1.01	1.43	1.76	1.96	2.16	2.55	3.20
Texas Utilities Fuel	0.78	1.30	1.49	1.56	1.70	2.08	2.92
Channel Industries Gas	1.37	1.25	2.09	2.21	2.27	2.54	3.14
Delhi Gas Pipeline Corp.	0.78	1.33	1.52	1.46	1.61	1.78	2.36
Intratex Gas	1.07	1.67	1.81	1.82	1.94	2.13	2.79
Houston Pipe Line	1.30	1.77	1.95	2.07	2.16	2.43	3.04
Weighted average	1.14	1.53	1.83	1.90	2.01	2.30	2.88
Interstate pipelines[b]							
Colorado Interstate Gas	0.35	0.47	0.71	0.81	1.10	1.56	2.12
El Paso Natural Gas	0.32	0.35	0.51	0.67	0.95	1.38	1.59
Michigan-Wisconsin Pipe Line	0.43	0.49	0.89	1.02	1.23	1.73	2.20
Natural Gas Pipeline Co. of America	0.33	0.41	0.72	0.83	1.17	1.62	1.88
Northern Natural Gas	0.30	0.38	0.59	0.73	0.95	1.17	1.54
Panhandle Eastern Pipeline	0.37	0.51	0.79	0.71	1.20	1.58	1.92
Tennessee Gas Transmission	0.35	0.42	0.67	0.88	1.19	1.59	2.11
Transcontinental Gas Pipeline	0.34	0.45	0.63	0.84	1.20	1.99	2.88
United Gas Pipeline	0.46	0.50	0.95	1.13	1.50	1.86	2.39
Weighted average	0.36	0.45	0.70	0.83	1.17	1.61	2.05

n.a. = not available.
a. Average cost for 1975–1981 is computed by dividing total purchased value by total volume purchased. Except for 1981, the data are taken from the annual reports of the Texas Railroad Commission. The costs for 1981 are based on data supplied by the Texas Railroad Commission. The listed pipeline companies are the largest in Texas.
b. The listed interstate pipelines are the principal interstate purchasers of natural gas in Texas. Average cost for 1975–1981 is computed by dividing total purchased-gas cost by (natural gas produced by the pipeline + gas purchased + natural gas withdrawn from storage). Data are taken from Federal Energy Regulatory Commission, form 2.

provides no guarantee that the data necessary for their application will be fully available, and they are not. The data do generally confirm the conclusions suggested by analysis of the NGPA's terms: intrastate pipelines are now at a disadvantage in bidding for deregulated gas, and that disadvantage will prob-

TABLE 2
AVERAGE PURCHASE COST OF GAS: COMPARISON OF INTRASTATE AND INTERSTATE PIPELINES, LOUISIANA, 1980 AND 1981
(dollars)

	1980	1981
Intrastate pipelines[a]		
Louisiana Intrastate Gas	2.64	3.44
Louisiana Resources	2.63	2.93
Monterey Pipeline	2.06	2.43
Sugar Bowl Gas	2.92	3.70
Weighted average	2.60	3.20
Interstate pipelines[b]		
Arkansas-Louisiana Gas	1.42	1.80
Louisiana-Nevada Transit	0.53	0.61
Mid-Louisiana Gas	1.63	2.07
Southern Natural Gas	1.69	2.33
Texas Eastern Transmission	1.58	1.87
Texas Gas Transmission	1.64	2.02
United Gas Pipeline	1.86	2.39
Weighted average	1.67	2.12

a. From "The Impact of the NGPA on Intrastate Markets in Louisiana," Submission of the State of Louisiana in FERC Docket No. RM82-26-000 (June 1982) (prepared by Foster Associates, Inc., Washington, D.C., for the Louisiana Office of Conservation).
b. From FERC form 2 (total purchased-gas cost − natural gas produced + natural gas purchased + natural gas withdrawn from underground storage).

ably increase in 1985. The precise extent of that disadvantage is, however, subject to considerable uncertainty.

The available data are of two kinds. One consists of average gas costs for intrastate and interstate pipelines. The other consists of projections of future natural gas prices and volumes by the Department of Energy's Office of Policy, Planning, and Analysis[68] and Energy Information Administration.[69] Average gas costs for Texas and Louisiana are set out in tables 1 and 2. The Texas intrastate pipelines listed in table 1 are the ten largest in the state; collectively they account for the overwhelming bulk of the pipeline segment of the Texas intrastate market. Comparable data for the Louisiana intrastates are not publicly available; the four intrastate pipelines listed in table 2 are the ones discussed in a recent Foster Associates report, and the data are taken from that report.[70]

The interstate pipelines in tables 1 and 2 are the principal interstate purchasers of gas in the two states. They are thus the pipelines that compete directly with the intrastate pipelines in the Texas and Louisiana field markets. Interstate pipelines may also compete with intrastates in burner-tip markets, notably through off-system sales; indeed, to the extent that sales are limited by demand rather than by supply, it may be in the competition for customers rather than in the competition for gas that the uneven distribution of the cushion is most important.[71] Burner-tip competition is not the principal concern of this paper, however, and interstate pipelines that may sell gas in Texas or Louisiana but are not major purchasers in those states are therefore not included in the tables.

In Texas the disparity in average gas costs between intrastate and interstate pipelines widened between 1975 and 1977, declined through 1980, and then widened again appreciably in 1981. No comparable data are available for Louisiana for the years before 1980, but in Louisiana also the difference in average gas costs increased substantially between 1980 and 1981. It would be hazardous to project a trend on the basis of two years' gas costs, but it can at least be said that the available cost data give no reason to believe that the disparity is becoming smaller.

To go beyond that very limited statement, it is necessary to turn from current cost data to studies projecting future cost and supply. Two studies provide a basis for estimating the future bidding disparity between interstate and intrastate pipelines: a November 1981 study by the Department of Energy's Office of Policy, Planning, and Analysis[72] and a December 1981 study by the department's Energy Information Administration.[73]

To estimate the interpipeline bidding disparity from these studies, it is necessary first to compute the interstate and intrastate natural gas cushions. The cushion is equal to the difference between the amount that pipelines pay for regulated gas and the amount they would pay for the same amount of gas under complete decontrol. Three alternative complete-decontrol prices are used. The high and mid-range prices are taken from the OPPA study.[74] In the light of subsequent market developments, however, that study's low price projection for complete decontrol does not now appear to be a realistic lower bound; the low complete-decontrol price used here, therefore, is substantially lower than the one projected in the OPPA study.[75]

The interstate cushion is much larger than the intrastate one throughout the 1985–1990 period and in all estimates (see table 3). By itself, however, the difference in the absolute size of the cushions indicates nothing about the bidding position of the interstate and intrastate pipelines.[76] Their relative bidding positions depend also on the relative size of the two systems and their relative need for deregulated gas. If the ultimate standard of reference is the distribution of natural gas under complete decontrol, something like the measure of need parity developed in this paper best incorporates these adjustments. Need parity is a measure of the relationship between (1) the size of the cushion and (2) the amount of deregulated gas the pipeline needs to give it the same

TABLE 3

INTERSTATE AND INTRASTATE CUSHIONS, 1985–1990

(billions of 1980 dollars)

	OPPA Estimate		EIA Estimate	
	Interstate	Intrastate	Interstate	Intrastate
High estimate				
1985	19.49	5.98	21.00	7.22
1987	—	—	20.83	4.57
1990	11.83	2.39	19.05	3.51
Mid-range estimate				
1985	16.89	4.91	18.41	6.15
1987	—	—	18.35	3.92
1990	10.16	1.95	16.47	2.91
Low estimate				
1985	4.72	0.02	6.73	1.30
1987	—	—	6.35	0.70
1990	2.61	0.06	5.29	0.34

SOURCES: OPPA Study; EIA Study.

total volume it would have under complete decontrol;[77] the latter amount is in turn equal to the difference between a pipeline's volume under complete decontrol and its projected volume of regulated gas under the NGPA.

The relative bidding position of the interstate and intrastate pipelines measured by need parity is set out in table 4. The 1985 figures for the mid-range price estimate and the OPPA cushion projection will be used here to illustrate the table's implications. According to the table, in 1985 the interstate pipelines can pay $2.40 more than the complete-decontrol price for the amount of deregulated gas needed to give them the same total volume they would have under complete decontrol; intrastate pipelines can pay only $1.31 above the complete-decontrol price for the analogous quantity of deregulated gas. The difference between the two amounts is $1.09. This number is unlikely, however, to be the difference in the price that interstate and intrastate pipelines actually pay for deregulated gas. Pipelines are likely to pay roughly similar prices for deregulated gas; there is no reason for a pipeline to pay more than the going price, and a pipeline that pays less will be able to acquire gas only where no other pipeline can compete for it. The bidding disparity projected in table 4 thus will probably be reflected not in different prices for deregulated gas but in a distribution of supply that is more favorable to interstate pipelines than the one that would exist under complete decontrol.

The measure of the interstate pipelines' bidding advantage derived from the OPPA and EIA projections should be regarded as a rough indication of their

TABLE 4
NEED PARITY, 1985 AND 1990
(dollars per thousand cubic feet)

	OPPA			EIA		
	Interstate	Intrastate	Difference	Interstate	Intrastate	Difference
High estimate						
1985	2.77	1.60	1.17	2.98	1.93	1.05
1990	1.16	0.45	0.71	1.88	0.65	1.23
Mid-range estimate						
1985	2.40	1.31	1.09	2.61	1.64	0.97
1990	1.00	0.36	0.64	1.62	0.54	1.08
Low estimate						
1985	0.67	0.01	0.66	0.96	0.35	0.61
1990	0.26	0.01	0.25	0.52	0.06	0.46

SOURCES: OPPA Study; EIA Study.

relative bidding position and not as a precise prediction. The underlying projections of price and supply are, like any projection of complex economic phenomena, subject to a wide margin of error; this is reflected in the use of three alternative complete-decontrol prices and also in the differences between the OPPA and EIA projections of the natural gas cushion. In addition, both OPPA and EIA divide the national gas market into only two parts: interstate and intrastate. The latter thus includes direct sales by producers as well as sales to intrastate pipelines, and it extends to Arkansas, Oklahoma, and southeastern New Mexico as well as Texas and Louisiana.[78] The average price of gas is significantly lower for this wider intrastate market than for the Texas and Louisiana intrastate pipelines.[79] Even if the OPPA and EIA projections were completely accurate, therefore, the measure of need parity derived from them would understate the interstate pipelines' bidding advantage relative to the intrastate pipelines of Texas and Louisiana.

The average cost data presented earlier in this section do not suffer from these shortcomings. There is no reason to doubt their substantial accuracy, and they are concerned solely with the Texas and Louisiana intrastate pipelines. Their shortcoming is a different one: they do not directly measure pipelines' relative bidding positions. If the standard of reference is the distribution of gas under complete decontrol, then relative bidding position depends on both the size of the cushion and the amount of gas needed to give a pipeline the same total volume it would have under complete decontrol. Average gas costs also depend on the size of the cushion, but in addition they depend on the amount (and price) of the deregulated gas that the pipeline has *in fact* acquired. Average gas costs thus reflect not only a pipeline's bidding position but also its bidding strategy. Two pipelines with equally large cushions will have different gas costs if one chooses to bid aggressively for gas in the current market and the other does not. Similarly, a pipeline with a very small cushion may hold its average gas costs down by purchasing little deregulated gas and allowing its reserves to decline.[80]

The precise extent of the interstate pipelines' present and future bidding advantage is thus uncertain. This quantitative uncertainty should not, however, be allowed to obscure the basic point: current average gas costs and future price and supply projections both confirm that intrastate pipelines are now at a substantial bidding disadvantage and will continue to be so at least until the end of the decade. And it is hard to see how it could be otherwise. The interaction of the NGPA with the pre-NGPA price history of the interstate and intrastate markets clearly favors the interstate pipelines with respect to both the price of regulated gas and the schedule for deregulation. There appears to be nothing in the law or elsewhere to offset this bias. The inevitable result is a bidding advantage for interstate pipelines both now and in the future; and to the extent that natural gas is allocated through competitive bidding, that advantage must affect the allocation of supply between interstate and intrastate pipelines.

Notes

1. 15 U.S.C. 3301–432.

2. 15 U.S.C. 717ff.

3. Phillips Petroleum Co. v. Wisconsin, 347 U.S. 672 (1954). The NGA's basic rate and certification sections (4–5 and 7) apply to "any natural gas company," a term that is defined in section 2(6) as "a person engaged in the transportation of natural gas in interstate commerce, or the sale in interstate commerce of such gas for resale."

4. Three stages can be distinguished: (1) an initial stage in which the Federal Power Commission sought to regulate producers as individual public utilities; (2) a second stage in which it regulated them on an area or national basis but continued to use accounting concepts derived from public utility regulation; and (3) a third stage in which it used discounted cash flow analysis. Even in the third stage, the commission did not escape from the essential circularity of cost-based rates in an increasing-cost industry: the price ceilings set by the FPC were based on past costs, but the ceilings in turn determined the volume of production and thus, in an increasing-cost industry, the level of future costs.

5. Compare Shell Oil Co. v. Federal Energy Regulatory Commission, 566 F.2d 536 (5th Cir. 1978), aff'd, 440 U.S. 192 (1979).

6. Section 1(c) of the NGA creates an exception for pipelines that are engaged in interstate commerce in a legal sense but operate only within a single state and are subject to state regulation. Such pipelines are called Hinshaw pipelines, after the sponsor of the law that created the exception.

7. In its efforts to attract natural gas to the interstate market, the FPC created certain exceptions to this rule, but they were of limited importance. They have now been

largely superseded by sections 311 and 312 of the NGPA.

8. The exceptions are the FERC's power to set incentive prices under section 107(c)(5) and its power to set new just and reasonable rates under sections 104, 106, and 109. The legislative history makes it clear that the commission is not required to use traditional cost-based analysis under the former provision, but the provision's logic evidently requires that some attention be paid to cost, since the provision is limited to gas production that presents "extraordinary risks or costs" and cost also determines whether a given incentive price is "necessary." The precise meaning of "just and reasonable" in the text of sections 104, 106, and 109 is currently a matter of debate, but presumably it at least includes, even if it is not limited to, traditional cost-based rate making.

9. For dedicated gas covered by NGPA sections 102(c), 103(c), and 107(c)(1)–(4), absolute dedication is replaced by a right of first refusal in the interstate pipeline (NGPA 315(b)). Absolute dedication continues in force for other interstate gas, principally old flowing gas covered by sections 104 and 106(a).

10. As discussed below, however, previous dedication *is* relevant to the question of deregulation of section 103 gas: section 103 gas not previously dedicated to interstate commerce is deregulated in 1985 or 1987; previously dedicated section 103 gas is never deregulated.

11. The NGPA uses several different dates that are linked somehow to the legislative process that produced the act. With respect to dedication to interstate commerce, two different dates are used: April 20, 1977, and November 8, 1978. The former is the date of President Carter's speech that formally began the legislative process, and the latter is the day before enactment of the NGPA.

12. NGPA section 121(a)(2) and (c). If, however, the new onshore well is in a reservoir not commercially produced before April 20, 1977, the gas is deregulated in 1985 regardless of whether it was previously dedicated to interstate commerce. Gas from the federal outer continental shelf is deregulated in 1985 only if it is produced from a lease entered into on or after April 20, 1977. Since all such gas is deemed to be in interstate commerce, this category has no precise counterpart in the intrastate market. The practical effect, however, of the limitation on deregulation of outer continental shelf gas is to increase interstate pipelines' relative share of the forever-regulated gas supply.

13. NGPA section 121(a)(2), (a)(3), and (c). For natural gas governed by existing or successor contracts, deregulation solely on the ground that it sells for more than $1.00 on December 31, 1984, is limited by section 121(e), where the price results from the operation of an indefinite price escalator clause.

14. Energy Information Administration, *The Current State of the Natural Gas Market: An Analysis of the Natural Gas Policy Act and Several Alternatives* (December 1981) (hereafter cited as *EIA Gas Market Study), p. 49.*

15. The following discussion of the implications of partial decontrol is based on relatively simple assumptions about pipeline bidding and is intended as an introduction to the bidding disparity problem. The significance of more complex, and perhaps more realistic, assumptions is discussed in the section "Pipeline Bidding Behavior" below.

16. Ultimately the constraint is the ability of the pipelines' distribution company customers to resell the gas in the burner-tip market at a price that covers the pipelines' purchased-gas costs plus a margin for transmission and distribution costs. (For direct pipeline sales, the constraint is of course the pipelines' own ability to sell gas to their

own end-user customers at a price that covers purchased-gas costs plus transmission costs.) The ability to resell the gas is not uniquely determined by purchased-gas costs; it is not possible, in other words, to represent the relationship between purchased-gas costs and burner-tip sales volume by a simple demand curve. To begin with, the average margin for distribution and transmission costs can vary, especially in the short run. That margin is determined by FERC regulation for pipelines and by state commissions for distribution companies. If the regulatory agencies reduce the margin equally for all pipelines or distribution companies, the short-run effect for natural gas producers and consumers would be the same as an equal relative increase in each pipeline's cushion: on the assumption of competitive bidding, the field-market price would be increased, and consumers would benefit only to the extent that the higher field price elicited additional supply. Similarly, if the margin were reduced only for some pipelines or distribution companies, the short-run effect would be the same as an unevenly distributed addition to the natural gas cushion: markets served by pipelines or distribution companies whose margins had been disproportionately reduced would enjoy some combination of larger supply and lower price. Of course, over the longer run the consequences might be quite different, as the smaller margin affected pipelines' or distribution companies' ability to serve.

The amount of gas sold also depends on rate design. For a given average purchased-gas cost it is possible to increase sales volume through price discrimination, charging a lower price to more price-elastic customers and a higher price to less price-elastic ones. See Robert Means, *A Survey of the Current Debate over Natural Gas Policy,* June 1981 (Report for Texas Energy and Natural Resources Advisory Council's Natural Gas Issues: Working Paper Series), pp. 36–38.

17. The shift in market shares as a result of the uneven distribution of the cushion can thus occur in two ways, which can be represented by the following schematic illustrations:

A. Shallow-cushion ⟶ Alternative fuels ⟶ Deep-cushion
      pipeline   pipeline

B. Shallow-cushion ⟶ Deep-cushion
      pipeline   pipeline

In A, shallow-cushion pipelines lose customers to alternative fuels while deep-cushion pipelines gain customers from such fuels. Aggregate consumption of natural gas and alternative fuels may be unchanged, but the distribution of both gas and alternative fuels between regions or pipeline service areas is distorted. In B, deep-cushion pipelines acquire customers formerly served by shallow-cushion pipelines. To the extent that this occurs, the bidding disparities affect the pipelines but do not distort the allocation of natural gas or alternative fuels. In reality, market shifts are likely to take both forms in those regions where interpipeline competition is possible. In such regions, shallow-cushion pipelines are likely to lose customers to both alternative fuels and deep-cushion pipelines. It is important to note that the "shifts" referred to here are shifts relative to the distribution of natural gas at equilibrium under complete decontrol or with an evenly distributed cushion. Thus a pipeline with a relatively shallow cushion may in fact expand its market if the starting point is one in which there is considerable opportunity for natural gas to displace high-valued alternative fuels. (Of course, such a pipeline could expand its market even more if it had a deeper cushion.)

18. This effect may be offset by the effect of the supply shift on fixed charges, which will of course increase per unit of sales as volume decreases. The relative importance of the two effects depends on the size of a pipeline's cushion in relation to the sum of its own fixed costs and those of its distribution company customers. If the cushion is larger, reductions in volume, if achieved by reducing purchases from high-cost sources, will cause average costs to decrease. If fixed costs are larger, reductions in volume will cause average costs to increase. If demand is sufficiently elastic, the latter case could theoretically degenerate into a "death spiral," in which there is no volume of sales for which demand is sufficient to support the full fixed and variable costs of the pipeline and its distribution companies. Presumably the end result of this spiral would be a reduction in fixed costs through corporate reorganization. More likely, however, the loss of some industrial markets would result in a considerable surplus of gas committed to the pipeline but not salable at current prices, and this surplus would in turn exert downward pressure on wellhead prices. It is not clear at what price equilibrium would be reached. It is plausible to assume, however, that the average wellhead price would be forced down far enough to allow at least some industrial markets to be served with gas.

19. FERC's sales jurisdiction is limited to sales for resale. See 15 U.S.C. 1(b). Interstate pipelines' direct sale customers nevertheless benefit indirectly from the pipelines' cushion. Under FERC rate-making practice, direct sales offset a pipeline's purchased-gas costs in an amount equal to the volume of gas involved in the direct sales times the pipeline's weighted average cost of gas. The offset reduces the amount that the pipeline is entitled to collect from its jurisdictional customers and thus, from the pipeline's standpoint, determines the cost of the direct sales. Since the cushion reduces a pipeline's weighted average gas cost, it also reduces the break-even point for direct sales.

20. See 15 U.S.C. 717(c). The Hinshaw exemption is primarily intended for local transportation companies, transporting gas for consumption within the state but receiving gas wholly or in part from out of state and thus in interstate commerce. Louisiana Power & Light Co. v. Federal Power Commission, 483 F.2d 623 (5th Cir. 1973), cert. denied 416 U.S. 974. Indeed, for some purposes Hinshaw pipelines are treated as local distribution companies. See NGPA, sections 2(15)–(17). To qualify for Hinshaw status under FERC regulations, a gas pipeline company's transportation activities must be confined to the boundaries of a single state, all gas transported on the system must be consumed in that state, and the company's rates and service must be regulated by a state commission (18 C.F.R. 152.1).

21. Faustina Pipe Line Company in Louisiana is an example (and probably the principal example) of this category. See Faustina Pipe Line Co., 16 FERC 61,128 (1981). Most of Faustina's supply is obtained from fields in Louisiana.

22. See text accompanying notes 37–42.

23. NGPA sections 311 and 312(b) permit intrastate pipelines to sell or assign gas to interstate pipelines or to local distribution companies served by interstate pipelines.

24. Energy Information Administration, Department of Energy, *Natural Gas Annual 1980,* February 1982, p. 13.

25. Energy Information Administration, *Intrastate and Interstate Supply Markets under the Natural Gas Policy Act,* October 1981, p. 9 (hereafter cited as *EIA Intrastate/*

Interstate Supply Study). EIA's information was based on information collected by the FERC for purposes of implementing sections 105 and 106(b) of the NGPA. Most Hinshaw pipelines and certain small producers were exempted from these reporting requirements.

26. In states served principally by interstate pipelines, many of the direct producer sales contracts found in the 1978 survey were probably, on the buyer's side, a response to interstate pipeline curtailments or threatened curtailments. See Department of Energy *Statistics of Interstate Natural Gas Pipe Line Companies* (published annually 1977–1980), showing national increases each year in the volumes transported for others, from 12.9 percent of total volume in 1977 to 15.7 percent in 1980. Curtailments have now largely disappeared, and the wellhead price for most of the uncommitted new supplies available for new direct sales is higher than the pipelines' average cost of gas.

27. EIA, *Natural Gas Annual 1980*, pp. 13 (production data), 20 (net interstate exports), 28 (consumption).

28. In 1978, the year the NGPA was passed, industrial consumption and electric utility consumption accounted for 86.90 percent of consumption in Texas and 90.86 percent in Louisiana. By comparison, the national average, excluding Texas and Louisiana, was 31.57 percent. Compare Energy Information Administration, Department of Energy *Energy Data Reports: Natural Gas Production and Consumption 1978*, October 1979 (state-by-state consumption data) with idem, *Energy Data Reports: Mainline Natural Gas Sales to Industrial Users 1978*, January 8, 1980 (state-by-state mainline industrial sales date). Only about 6 percent of the industrial demand in Texas and in Louisiana was met by mainline sales by interstate pipelines.

29. See table 1 in this chapter.

30. The states are given in order of decreasing net export volumes for 1980. EIA, *Natural Gas Annual 1980*, p. 20.

31. In 1978 combined industrial and electric demand amounted to 66.24 percent of total consumption in New Mexico, 80.42 percent in Oklahoma, and 61.65 percent in Kansas. EIA, *Natural Gas Production and Consumption 1978*, p. 8 (computed percentages). For Texas and Louisiana, see note 28.

32. Throughout the 1960s and 1970s, intrastate prices were higher than interstate ones nationwide. *EIA Intrastate/Interstate Supply Study*, p. 40. These national averages were, however, dominated by Texas and Louisiana, which accounted for about 72 percent of the national intrastate market in 1978 (ibid., p. 10). In that year the average intrastate price was higher than the interstate one in all five states, but for intrastate pipelines it was much higher in Texas and Louisiana than in New Mexico, Oklahoma, and Kansas. The average cost of gas purchased by intrastate pipelines and the overall average cost (in dollars) of intrastate gas, including direct sales, for the five states in 1978 were as follows:

	Pipelines	Total
Texas	1.71	1.40
Louisiana	1.62	1.12
New Mexico	1.38	1.38
Oklahoma	1.31	1.30
Kansas	1.17	1.16

Ibid. (figures converted on basis of 1.00 thousand cubic feet = 1.02 million Btu).

In some respects, however, prices in new and renegotiated contracts are a better measure of current market conditions than average prices, which may also reflect prices determined by older contracts that are not responsive to current market conditions. From the third quarter of 1975 until the first quarter of 1977, the Federal Power Commission tracked the prices paid in new and renegotiated nonjurisdictional contracts of interstate companies. According to the last of these reports, issued on May 17, 1977, the average prices (in dollars) for these contracts were as follows:

	New	Renegotiated
Texas	1.9443	1.9785
Louisiana	1.5490	1.3787
New Mexico	—	1.3785
Oklahoma	1.6362	1.6686
Kansas	1.4526	—

Intrastate Natural Gas Prices of FPC, Jurisdictional Natural Gas Companies Selling More Than One Million MCF per Year in Interstate Commerce: Summary by State and FPC Gas Pricing Area, January, February, March 1977, Bureau of Natural Gas Staff Report (Washington, D.C., May, 1977). These statistics on new contracts suggest that Texas might more properly be treated as sui generis.

33. See N. Mex. Stat. Ann. 62-7-1ff.; 52 Okla. Stat. Ann. 260.1ff.; Kan. Stat. Ann. 55-1401ff. The Kansas and Oklahoma statutes apply to intrastate contracts made before April 20, 1977. Kan. Stat. Ann. 55-1403; 52 Okla. Stat. Ann. 260.3. Under these statutes, indefinite price escalator clauses in such contracts are allowed to escalate prices only to section 109(b) levels. Kan. Stat. Ann. 55-1405; 52 Okla. Stat. Ann. 260.5. Gas from new wells is treated as if it had been under interstate contracts before the NGPA. Kan. Stat. Ann. 55-1406; 52 Okla. Stat. Ann. 260.6. Renegotiation is permitted, Kan. Stat. Ann. 55-1408; 52 Okla. Stat. Ann. 260.8. The restrictions are removed on December 31, 1984. Kan. Stat. Ann. 55-1411; 52 Okla. Stat. Ann. 260.13. When the NGPA was enacted, New Mexico already had a statute governing the pricing of wellhead sales for resale in intrastate commerce, and the New Mexico legislature's response to the NGPA took the form of two amendments to the existing statute, one in 1979 and one in 1981. As it stands now, the law applies to all sales in intrastate commerce for resale and consumption in New Mexico unless commingled with gas destined for out-of-state consumption. N. Mex. Stat. Ann. 62-7-4. In general, such gas is subject to a formula based on a July 1981 price of $1.62 per thousand cubic feet and the consumer price index. N. Mex. Stat. Ann. 62-7-6(A). Five categories of gas are excepted from the formula. Gas from stripper wells is entitled to NGPA pricing as if it had been dedicated to interstate commerce, N. Mex. Stat. Ann. 62-7-6(B). Gas from wells commenced during 1973 and 1974 get similar treatment. N. Mex. Stat. Ann. 62-7-6(c). State-owned production sold under rollover contracts gets the NGPA section 106(b)(2) price. N. Mex. Stat. Ann. 62-7-6(D). High-cost gas is subject to incentive pricing by the state commission. N. Mex. Stat. Ann. 62-7-6.1. Finally, gas sold from wells completed after July 1, 1981, to state-regulated and municipally owned public utilities and to intrastate pipelines is exempted from the general ceilings if a number of conditions are met, including a minimum contract term of fifteen years. For the exemption to apply, the purchaser of gas from such a new well must also believe that "the producer's interest in the volume of estimated recoverable reserves from the new well is equivalent or greater than the producer's interest in the volume of estimated remaining recoverable reserves from the well producing natural gas," which would otherwise be

101

subject to the general pricing provision. N. Mex. Stat. Ann. 62-7-6.2. The New Mexico law will expire on July 31, 1983, if the NGPA schedule for wellhead price decontrol is not changed, N. Mex. Stat. Ann. 62-7-9. It may expire earlier if total deregulation is accelerated and may be reinstated if new controls are imposted. Ibid.

34. Energy Reserves Group, Inc. v. Kansas Power & Light Co., 230 Kan. 176, 630 P.2d. 1142, prob. juris. noted, 456 U.S. 904 (1982).

35. Intrastate sales in 1978 were 225 billion cubic feet and 143 billion cubic feet for California and Colorado respectively, compared with 121 billion cubic feet for New Mexico. *EIA Intrastate/Interstate Supply Study,* p. 10.

36. EIA's survey lists no sales to intrastate pipelines for California in 1978; all the sales are direct producer–end user sales. See *EIA Intrastate/Interstate Supply Study,* p. 10. Most of California's intrastate gas movement occurs on the state's several Hinshaw pipelines (Southern California Gas Company, San Diego Gas and Electric Company, Pacific Gas and Electric Company, Southwest Gas Corporation, and CP National Corporation) or on local distribution systems that extend their own pipelines into the field.

37. *EIA Intrastate/Interstate Supply Study,* p. 10. Direct sales have also been made to end users served by interstate pipelines, under the commission's self-help program. See *Policy re Certification of Pipeline Transportation Agreements,* 54 FPC 821 (1975) (Order No. 533). But such direct producer sales are an insignificant factor in the interstate market.

38. Nationally, in 1978, 71.06 percent of the gas involved in direct intrastate sales was covered by contracts antedating 1970.

39. For 1978, average pipeline and direct sale prices were $1.71 and $0.59 respectively for the Texas intrastate market and $1.62 and $0.66 respectively for the Louisiana intrastate market. *EIA Intrastate/Interstate Supply Study,* p. 10 (prices stated in 1978 dollars and converted from caloric to volumetric basis on the basis of 1.00 thousand cubic feet = 1.02 million Btu).

40. For interstate pipelines this practice is generally mandated by the FERC. No similar legal requirement appears to apply to the industrial sales that constitute the greater part of the business of Texas and Louisiana intrastate pipelines, but their sales contracts appear commonly to base price on average cost.

41. If the firm's rights under a low-cost contract are not transferable, the firm can obtain the rent associated with the low price only by remaining in business, perhaps only by remaining in business at a certain location. Its ability to remain in business at that location may then depend on the average of the cost of gas under the old contract and the cost of newer supplies. In this sense, an unregulated firm may engage in a kind of internal rolled-in pricing.

For intrastate pipelines' unregulated industrial sales, pricing based on average rather than marginal cost appears to be explained by two considerations. First, rolled-in pricing may be necessary if an intrastate pipeline is to meet the competition of alternative fuels while covering its own costs; in this respect intrastate pipelines are like the industrial firm that uses internal rolled-in pricing because this is the only way that it can benefit from its low-cost sources of supply. Second, rolled-in pricing also serves as a device for allocating the risk of price fluctuations between intrastate pipelines and their customers.

42. See *EIA Intrastate/Interstate Supply Study,* p. 11.

43. The counterassumption discussed here is that the average field price is determined by pipeline bidding but that the bidding is to some extent noncompetitive and therefore stops short of the market-clearing level. An alternative counterassumption is that the average field price is determined by the price escalator clauses in existing contracts and not by current bidding. That counterassumption implies that the average field price could (but not that it necessarily will) exceed the market-clearing level.

44. It is necessary to refer to purchased-gas cost rather than to total burner-tip price because the latter also includes transmission costs, which vary with distance from the natural gas field. As a result of the lower transmission cost of intrastate pipelines, the total price to their customers may be lower than the burner-tip price to customers of interstate pipelines even if the purchased-gas-cost component of the price is higher.

45. See text at notes 52–57.

46. It is necessary to speak of relative share rather than absolute amount because the total amount of natural gas will tend to be different under partial decontrol from what it would be under complete decontrol.

47. A $10 billion cushion embodied in 6 Tcf of price-controlled gas implies that the average price of that gas is $2.33, that is, $1.67 less than the complete-control price of $4.00. The 6 Tcf of uncontrolled gas therefore must be bid up to $5.67 to give a market-clearing average price of $4.00. On similar reasoning, the average price of the intrastate pipelines' price-controlled gas is also $2.33, and a $5.67 price for 3 Tcf of uncontrolled gas will also give those pipelines an average price of $4.00 for their total supply of 6 Tcf.

48. As discussed earlier, comparison with a reference point defined by the interstate and intrastate pipelines' relative pre-NGPA bidding positions would probably show the intrastate pipelines at a more serious disadvantage than one based on the complete-decontrol reference case. Some policy discussion, however, appears implicitly to define its reference point in terms of burner-tip prices: intrastate pipelines are not considered improperly disadvantaged if their burner-tip prices are no higher on average than burner-tip prices in the interstate market. This reference point in effect accepts the legitimacy of an interstate field-market bidding advantage as long as that advantage does no more than offset the intrastate pipelines' lower transportation costs.

49. See Energy Information Administration, *An Analysis of Post-NGPA Interstate Pipeline Wellhead Purchases* (1982) (hereafter cited as *EIA Post-NGPA Wellhead Purchases Study*). According to this study, the only major pipeline company that had not made any purchases of high-cost section 107 gas by August 1982 was Consolidated Gas Corporation, and Consolidated acquires most of its gas from pipelines that do purchase high-cost gas (ibid., pp. 28, 74). For the other nineteen companies, the average cost of high-cost gas ranged from $4.76 to $8.09 per thousand cubic feet (Mcf) in mid-1982. By contrast, the average pipeline sales price was $3.75/Mcf to industrial customers and $3.82/Mcf to wholesale customers. Monthly Gas Industry Activity Report No. 27, prepared by the Office of Regulatory Analysis, Federal Energy Regulatory Commission (November 30, 1982). A comprehensive survey of all companies comparing their costs of gas with their rates is unavailable. It appears, however, that Northern Natural Gas, which spends the least for high-cost gas ($4.76/Mcf) of the nineteen major companies reporting such purchases (ibid.), nevertheless spends more than the highest rate it charges to customers for emergency purchases ($3.85/Mcf). Natural Gas Pipe Line Company, which at $5.15/Mcf has the next least costly supply of

high-cost gas, has a maximum commodity charge of $3.62/Mcf.

50. American Gas Association, *A Statistical Analysis of Bidding Trends for Decontrolled Natural Gas under the NGPA* (March 19, 1982); *EIA Gas Market Study* (December 1981); J. Price and Saida Shaalan, *Pipeline Behavior in Acquiring NGPA Section 107 Gas* (July 1982).

51. There is a strong inverse correlation between pipelines' volumes of old gas and the average price of that gas. In other words, pipelines with relatively large volumes of old gas tend also to have a relatively low average cost for such gas. In mid-1982 Texas Eastern, Natural, and Florida Gas Transmission purchased the largest proportional amounts of old gas among the twenty largest pipeline companies—81.03 percent, 70.67 percent, and 69.49 percent, respectively. Their average cost of gas ranked first, third, and fourth lowest—$1.34/Mcf, $1.92/Mcf, and $1.92/Mcf, respectively. At the other end of the spectrum, Consolidated, Transcontinental, and Columbia had the lowest proportion (eighteenth, nineteenth, and twentieth respectively) of old gas—27.45 percent, 41.30 percent, and 41.39 percent. They were among those with the most expensive regulated gas: Consolidated (sixteenth, $2.58/Mcf); Transcontinental (twentieth; $2.90/Mcf); and Columbia (nineteenth; $2.75/Mcf). See EIA Post-NGPA Pipeline Wellhead Purchases Study, pp. 62–142 (computed percentages and rank). More generally, statistical analysis for the twenty-company sample of mid-1982 shows a significant negative correlation ($r = -0.682$) between the percentage volumes of old gas purchased and the average cost of the old gas.

52. As a variant on this strategy, a pipeline faced with a declining load may use its cushion to underprice other pipelines in making off-system sales. See Office of Regulatory Analysis, Federal Energy Regulatory Commission, *Off-System Sales: A Preliminary Outline of the Policy Issues* (1982), submitted as comments in Review of Off-System Sales Program, FERC Docket No. GP82-47-000 (October 12, 1982).

53. The intertemporal bidding assumption is based in substantial part on current modeling work being done by the Energy Information Administration. EIA bears no responsibility, however, for either the particular formulation given to the assumption here or the discussion of its implications.

54. The link between present and future commitment of the cushion assumes some institutional rigidity in the negotiation of contracts between pipelines and producers. In theory a contract might provide for a high present price and a lower future one, so that full current commitment of the cushion would be consistent with additional purchases of deregulated gas in the future. It is possible that the close link between present and future commitment could also be eliminated through the use of market-out clauses, but the practical implications of such clauses are still uncertain.

55. At the limit, this strategy would involve a ''cheapest-first'' policy: within the constraints imposed by problems of take-or-pay, drainage, and similar considerations, a pipeline would take all available gas at a certain price before taking any gas at a higher price. Anything short of a cheapest-first policy would in theory leave the pipeline with some flexibility to increase its purchase of new deregulated supplies by increasing the proportion of low-cost gas in its current take.

56. There appear to be two potential economic reasons for a pipeline to choose to defer use of its cushion: (1) it may fear that if it fully committed its cushion now through the purchase of deregulated gas, it would not be able to continue to resell that gas in the future when its cushion was smaller; or (2) it may believe that deregulated gas can be

acquired more cheaply in the future. These potential economic reasons are to be distinguished from an unwillingness to bid aggressively for deregulated gas out of a concern for possible political or regulatory consequences.

57. It may use it in the future either to bid for then uncommitted gas or to offset generous deregulation clauses that it used for nonprice bidding in the earlier market. Note, however, that there *is* some loss in the deferred commitment strategy, since that strategy, unlike the deferred production strategy, does use up the cushion in a physical sense: less of the inexpensive gas will remain to be produced in the future since it is being produced today.

58. The following discussion is concerned solely with ambiguities in the concept of a short-term market-clearing price. Additional ambiguities arise if a longer time perspective is used. Demand for natural gas varies both seasonally and with changes in the general level of economic activity, while natural gas field prices are relatively rigid. It is therefore virtually inevitable that short-term imbalances between supply and demand will occur at some times, and indeed it is likely that an imbalance in one direction or the other will exist most of the time. See Robert Means, "Issues in the Debate over Natural Gas Decontrol," *Public Utilities Fortnightly*, vol. 18 (October 28, 1982). If the price rigidity is taken as given, it may be more reasonable to focus on a longer time perspective. For example, pipelines may in fact behave as if demand exceeded supply so long as supply is not adequate to meet their peak demand during a period of normal economic activity and therefore may bid aggressively for additional reserves even when supply clearly exceeds demand in the short term.

59. See note 16.

60. Ambiguity regarding the amount of demand stems from the continued existence of nonprice constraints on natural gas consumption. Such constraints do not now appear to be quantitatively significant.

61. The latter amount is the amount that TGP must purchase from associated gas wells and wells subject to drainage and from wells that would suffer a permanent loss of deliverability if takes were not maintained.

62. In mid-1981, according to filings made in connection with applications for off-system sales, at least seven major gas pipelines were projecting prepayments under take-or-pay contracts. Notice of Informal Public Conference, Review of Off-System Sales Program, FERC Docket No. GP82-47-000 (August 6, 1982). Systematic data for 1982 are not yet available, but an informal survey of recent rate filings indicates that at least several additional pipelines have made prepayments.

63. Four pipelines were then purchasing section 107 gas at an annual rate of less than 0.5 billion cubic feet, while five others were paying less than $5.00 per thousand cubic feet (in January 1982 dollars). *EIA Post-NGPA Wellhead Purchases Study*, p. 28. The latter pipelines presumably were confining their section 107 purchases to regulated tight sands gas or to deregulated gas for which there was relatively little competition.

64. The exception was Consolidated, which purchases about 80 percent of its gas from other pipelines (ibid., pp. 28, 74).

65. Only one pipeline was paying an average price of less than $5.00, and only three were paying an average price of less than $5.50 (in January 1982 dollars).

66. Uncommitted natural gas falls into two general categories. The first is gas discovered on acreage not previously committed to a pipeline. Such gas is generally governed by section 102 or 107, and most of it is deregulated in 1985. The second

category consists of natural gas that was subject to a contract that has now expired *and* that was not dedicated to interstate commerce. (Dedicated gas must continue to be sold to the same interstate pipeline after the initial contract expires unless the gas is classified under section 102(c), 103, or 107(c) (1)–(4). See NGPA, section 315 (b).) Such gas is governed generally by section 106 (b), and most of it will be deregulated in 1985. Compare NGPA section 121(e), which implicitly excludes rollover contracts from its limitation on deregulation under section 121(a)(3).

67. Indeed, in 1985 virtually all pipelines will find themselves committed to purchasing large amounts of deregulated gas regardless of the bidding strategy that they choose, since nearly all the section 102 gas they have been purchasing since 1978 will then be deregulated.

68. Department of Energy, Office of Policy, Planning, and Analysis, *A Study of Alternatives to the Natural Gas Policy Act of 1978*, November 1981 (hereafter cited as OPPA Study).

69. Department of Energy, Energy Information Administration, "An Analysis of the Natural Gas Policy Act and Several Alternatives, Part I," December 1981 (hereafter cited as EIA Study).

70. "The Impact of the NGPA on Intrastate Markets in Louisiana," Submission of the State of Louisiana in FERC Docket No. RM82-26-000 (June 1982) (prepared by Foster Associates, Inc., Washington, D.C., for the Louisiana Office of Conservation). A notable omission from table 2 is Texaco, which is still the largest intrastate pipeline in Louisiana. Much of Texaco's gas is sold at a relatively low price under long-term contracts, and inclusion of this gas at the contract price would substantially reduce the intrastate average for Louisiana. Nearly all of the gas sold by Texaco, however, is from its own production. The low-price contracts are dedicated to particular customers and in general are not being renewed at their expiration. Texaco's pipeline sales thus appear to be functionally similar to direct producer sales, and the low-price contracts do not appear to provide an economic basis for current field market bidding (ibid., pp. 43–49).

71. Burner-tip competition may, of course, indirectly affect the field market. To the extent that intrastate pipelines lose customers in this burner-tip competition, their ability to bid for new reserves will presumably be diminished.

72. OPPA Study.

73. EIA Study.

74. OPPA Study, app. A, attachment 4 (Summary output tables "REFFUL82" and "HIGH-FUL82"). The December 1981 EIA study did not examine the complete-decontrol alternative.

75. OPPA high and mid-range complete-decontrol prices and the low complete-decontrol price used here are (in 1980 dollars):

	High	Mid-Range	Low
1985	5.09	4.65	2.65
1987	5.36	4.92	2.76
1990	6.08	5.50	2.93

The low price was computed on the assumption that the complete-decontrol price would be equal to 70 percent of the Btu-equivalent price of oil and that oil prices would increase at the rate of 2 percent annually from a 1985 level of $22.00.

76. The intrastate cushion estimates in fact include direct producer sales as well as sales to intrastate pipelines.

77. For this purpose imported gas is treated as deregulated gas and is assumed to sell at the same price as deregulated gas.

78. For both studies the intrastate market consists basically of intrastate sales in the Department of Energy's Region IV.

79. The average price of gas in direct producer sales in Texas and Louisiana is significantly lower than the average price for sales to intrastate pipelines in those states, and the intrastate price is also higher in those states than in the other principal states in the intrastate market.

80. See "Impact of the NGPA in Louisiana" (Foster Associates), pp. 13–18, indicating that reserves committed to Louisiana intrastate pipelines have declined since passage of the NGPA.

Commentary

Jack Earnest

I will start by saying that I think there is in effect rationing of natural gas, not in the classic sense of curtailments, although to a very limited degree that could happen today. I do not think it is happening except where induced by the weather, but there are some institutional restraints. Neither Transcontinental nor Texas Eastern has the freedom to select new customers. That is a point I think we often forget when we talk about a market; there is no opportunity to sell to new customers. This creates problems for any pipeline. Quite obviously, Transco and Texas Eastern both have contracts with customers, principally distributing companies, that call for certain volumes of gas. There is no mechanism for them to increase those except by seeking certificate authority to make additional sales. That gets to be a severe problem in a market-oriented situation. To that degree, then, I would say there is rationing, and it helps explain why practices differ from what would be expected in pure economic theory.

Furthermore, from a pipeline's standpoint, there is an absolute service obligation—at least that is asserted by some—to provide contract volumes; therefore, a pipeline that has an extremely low cushion does not have the freedom of choice to say, "I will not buy from this gas supply to meet my market." The alternative that pipeline may be facing is litigation, damage claims for failure to serve the market at whatever cost. Such damage claims may very well be based not on the cost of the gas or the delivered price of the gas that was not delivered but on alternative fuel at much higher cost.

I think that Robert Means has an excellent point: that it is very difficult to talk about the intrastate pipeline market or supply versus the interstate pipeline market or supply. It truly is a pipeline-to-pipeline situation. Each pipeline has the option of not buying; in a perfect world, its decision would be made rationally on the basis of whether the gas was resellable in its particular market.

There is not just one market for natural gas, there are several markets. A Texas intrastate pipeline system that has been in existence for a long time may have built its pipeline largely to serve an industrial market load—that is typical of some of the older pipelines in Texas. Some of that market load has disappeared, probably forever, because a particular industry can no longer use natural gas and remain competitive with others located elsewhere and using a different gas supply. That is an unfortunate fact of life. The same thing happens to some interstate pipeline systems. Today there are several interstate pipelines
108

with severe marketing problems; they have extensive gas supplies that they cannot move through to market because of both the institutional restrictions on taking on new customers and the restrictions that arise because their markets just do not want natural gas at today's prices or have shut down their plants for other economic reasons.

There are many reasons why there is a temporary surplus today and why it is not evenly divided among the pipelines. The surplus is very bad for some pipelines, and other pipelines are very short. Transwestern, which goes to the West Coast, has about a four-year supply of gas and not much of a cushion. This suggests to me that it has two options. It can either, rather quickly, reach a point where it does not supply gas to the California market in accordance with its contractual and certificate obligations, or it must buy gas to supply those markets. When it starts to buy gas to supply its markets, it must compete with other pipelines; so the option is to meet the price or fail to supply the market. That is the decision each pipeline faces.

As for redesigning the system to provide advantage or disadvantage to a particular pipeline or group of pipelines, it may well be that national policy with all the wisdom that could be brought to bear might decide that gas burned under an electric utility boiler in Texas is not the right thing or that gas burned under an electric utility boiler in New York City is not the right thing. These decisions seem to me very difficult to adopt as a matter of national policy.

When all the pros and cons are weighed, there are two alternatives. One is a system that allocates gas supply by use, by area, and by region and restricts production by producer, by area, and by region, so that it all meshes and some long-term gas supply is maintained. Although the Soviets may be giving us a hard time on the building of new gas lines into Europe, they have tried this rationalization process in their gas industry and have had some amazing failures in building large pipelines that would not work because they had no compression equipment. They finally get it worked out, but with tremendous inefficiency.

The other alternative clearly is to let the chips fall where they may but try to eliminate all the artificial barriers. Over time—and it may not work out to everyone's satisfaction—the system will be rationalized more along economic lines. I do not know how to get from the absolute planned to the unplanned and hope that economic laws will make it work. The institutional barriers can be removed to some degree, although the NGPA may not take care of all the institutional problems. No one could write legislation that would take care of all of them, but at least we ought to start trying to eliminate the major barriers.

Robert Leone

I will focus my attention strictly on the market-ordering question. As I pursue the market-ordering problem, I feel a bit like the dog chasing the automobile: I'm not exactly sure what I'm going to do when I catch it. I am not entirely sure what it is we are supposed to do once we discover that we have a serious market-ordering problem. But I think my analogy with the dog applies because the issue of what the dog should do when he catches the car is more important in concept than in practice. I would like to suggest several reasons why I conclude that the market-ordering problem may be substantially less significant than our authors have suggested.

First, I would like to state what I see as the three principal lessons of these two papers. Two of the lessons have public policy implications, and the third is by way of a public acknowledgment of my own ignorance. In reading these papers, I have concluded that the length of time it takes me to understand any given set of natural gas pricing rules is longer than the period of time between major changes in those rules. That is the most persuasive argument for deregulation that I have heard in a long time.

The second lesson is that a major issue associated with natural gas decontrol is income redistribution, which relates to the problems of owners of new and old gas and of new and old pipelines. Politically these redistribution issues may be more important than the redistribution questions between producers and consumers. As I read these kinds of papers and see the incredible morass of regulation and procedure, I think more attention needs to be paid to these issues.

The third lesson in these papers is based on a premise that I believe but a logic that I am not sure I can extend. The premise is that there is a natural tendency in competitive markets to dissipate quasi rents. The whole notion behind this problem of cushions, or quasi rents in economic jargon, is that tracing the implications of dissipating these cushions in the single dimension of higher deregulated natural gas prices leads one to believe there is a problem of market disorder. There is no doubt in my mind that the basic premise is correct: competitive forces do serve the beneficial social function of dissipating quasi rents. But I would caution the readers of these papers that the arithmetic extrapolation of this proposition to a single dimension is very likely to overstate the market disorder effect.

At least as I look at the market-ordering problem in the unregulated and regulated natural gas markets, my sense is that it falls into the general category of problems in which expectations exceed the actual event. I think both Robert Means and Catherine Abbott have understated the creativity of gas utility employees, of gas utility customers, of regulators and legislators, and of gas utility managers in finding clever ways to dissipate these quasi rents. There are other factor markets, for example, in which the gas utilities who have this so-called cushion operate: there are clearly labor markets; there are clearly capital markets. There are classic experiences in other regulated industries where price advantages are dissipated in higher service costs, excess maintenance, higher customer hookup prices, cost sharing on industry boiler configuration changes, or subsidization of conservation activity. I can think of all kinds of ways in which a gas utility confronting a cushion would find it desirable to try to dissipate these rents. That is not to say that the problems of a cushion are irrelevant. It *is* to say that one does not necessarily find the problem of a cushion by looking at the natural gas price marketing problem.

To make the point, let me give a simple hypothetical example. Suppose that I am a vertically integrated gas utility. I happen to own a pipeline and some natural gas wells; there is a cushion, and there are very substantial prices for natural gas on the margin. Because I am operating in a competitive market situation, I am able to sell that natural gas at that very high price on the outside. Would I as a competitor prefer to sell the natural gas inside the firm or outside the firm at the equivalent price? Given that the short-run elasticities of demand tend to be different from the long-run elasticities and that I am not enthusiastic about triggering the long-run demand elasticities of my customers, if I faced equivalent prices both inside and outside the firm, I would have a strong incentive to sell the natural gas on the outside. If I flip that argument around and say, "what am I willing to pay as a purchaser of natural gas if I am a gas utility?" I come to the conclusion that I am not necessarily willing to pay all the price increases that I could expect to pass on in the short run in a competitive marketplace subject to the regulations in which I confront it.

I think it important to consider the policy implication of this market-ordering problem. The argument states that if there is a market-ordering problem with the present price path for natural gas, we should use the existence of that problem to justify a change in policy. I have no problem with that logic except for my concern that we may empirically observe far less of a market-disordering problem than we expect; having based arguments about decontrol on the existence of a market-disordering problem, we may incorrectly conclude that in the absence of such a problem continued regulation in this market is harmless. It is very dangerous to come to that conclusion because we simply do not find that the inefficiencies in continued policies of price control will manifest themselves in any single market distortion. We can expect numerous market distortions to result from policies of artificially constrained prices, and we should therefore not focus undue attention on a potential market-disordering problem that is likely to be more significant in concept than in practice.

Discussion

BENJAMIN ZYCHER: I don't think I understand Robert Leone's quasi rent argument, but my remarks may illustrate where I am confused. The process of bidding for decontrolled gas, given the presence of a cushion, is in fact a way of dissipating the cushion, dissipating the quasi rents. If the hypotheses that have been presented here are correct, and I believe they are, there is a way of shifting the quasi rents from consumers, who are the nominal beneficiaries of the price controls, to some producers, who will be the actual beneficiaries. In most competitive markets it is likely that bidding will take the form of cash offers. That is the utility-maximizing way of dealing with the quasi rent problem. So even if we accept the hypothesis that the quasi rents will be dissipated, it still leads to the conclusion that the form that the dissipation will take in a competitive market is that of high prices for decontrolled gas. The example of the vertically integrated utilities is a bit misleading because it abstracts from the competition for decontrolled gas on both sides of the market, in which the demand for gas is derived from the ultimate demand for gas by consumers who are willing to pay something like the oil-equivalent price.

MR. LEONE: There is a competitive tendency to dissipate the rents. The papers make the correct observation that deregulated natural gas is one outlet for that dissipation but then do arithmetic as if it were the only outlet. All that I am suggesting is that there are numerous outlets through which these rents can be dissipated; if we predicate our policy on looking at the empirical evidence in the single source of distortion and do not see that single source being as distorting as we expected because it has been spread about in a number of other distortions, we may illogically conclude something about the policy.

BENJAMIN SCHLESINGER: I think Robert Leone is suggesting that there is a kind of one-dimensionality to the thesis that pipelines will bid to market-clearing levels. That is not happening now for section 107 gas. There most certainly is a strong inverse correlation between pipelines' old gas holdings and their bidding on section 107 gas, as shown in the EIA data Catherine Abbott cited. I am not suggesting that high 107 prices would be sustained if there were no cushion. I concur with the thesis that the cushion is an enabling feature, but I believe there are other factors involved. I identify, therefore, with Robert

Leone's statement that there may be other rent captures waiting in the wings to make it impossible for pipelines to pay the average price of gas.

I would like to comment on the three strategies for contracts. I want to take them in reverse. Third was the containment strategy: let us set up limits to one contract escalator clause's triggering another. Everybody can figure out ways around that, and it will not work. The second, the outright ban on escalator clauses, has been suggested by the American Gas Association. I think that is an appropriate starting point for dealing with them. The first strategy was the cap idea of the Interstate Natural Gas Association (INGAA), which I think has been seriously misinterpreted. INGAA did not propose a permanent 70 percent cap on escalator clause gas prices. It proposed a 70 percent cap in a climate where that is appropriate, residual prices being what they are. It left open the prospect of redeveloping that measure or resetting the cap. There was not a great deal of precision in that aspect of the proposal, but the intent was to set an appropriate cap, perhaps even one based on market levels. The danger here is the danger of wholesale gas price increases: suddenly, for no reason other than that he did it, I am going to do it. Many people who have examined the behavior of the gas market in the early 1950s suggest that this problem may well have contributed to the imposition of price regulations in the first place. It is more serious than you all are giving it credit for.

EDWARD GRENIER: We have noticed in at least one purchased-gas-adjustment case that it is difficult to pin down the behavior of that pipeline to any one of Catherine Abbott's hypotheses. We have found out, first, that a pipeline that does not need section 107 gas now will not need it in the mid-1980s. The defense against paying too high a price is that there are good market-out clauses. So one wonders if they are building long-term reserves. They say, however, "Since we can walk away from them, how long-term are they? Besides, some of them will be dissipated." The flow rates will reduce the benefit of them; they will have to use them up now, and they are under high take-or-pay anyway. So one wonders what the strategy is. It does not really fit sensibly into any of these hypotheses. That is one observation.

On the subject of the contract clauses or the contract approaches, I think undue reliance on ideology can get in the way of moving to a free market. If we worry too much about using the regulatory process or controls to smooth the transition, we may end up with more chaos than we anticipate. What appeals to me in the INGAA approach is its simplicity. That is an important factor. We have been going through the exercise of looking for a statutory most-favored-nation clause, and that is so incredibly complicated it is like a piece of Swiss cheese. The one thing that approach just flatly overlooks is that the ingenuity of man has already devised a very high definite escalator provision—nothing to do with fancy indefinites, just a definite provision that, along with other escalators, escalates at 1 percent a month compounded forever. That sets a new floor every month. The INGAA approach hits that; the other approaches do not really contain that sort of thing.

One variation that is promising is, instead of sticking with a 70 percent cap, trying to peg the "market-clearing" price at the prices actually paid for new gas. There would be some lag, but the government could collect that data—no judgment required, just collect the prices paid for new gas contracts. It strikes me as a very good idea, so that we would not be tied to an exogenous price such as oil.

In response to Jack Earnest, I doubt that the certificate proceedings at the FERC, particularly with its current management, will really restrict anyone very much in getting new customers, especially with the possibility of temporary certificates. From the point of view of industrial consumers, we like at least some little check on whether pipelines are adding too much load too quickly in relationship to their reserves.

MR. EARNEST: I did not mean to imply that it is just the certification procedure; there is also the cause of the certification procedure. The real problem is this: If I have a five-year gas supply and, even though I have excess deliverability— more gas being shoved at me at this end of my pipeline than coming out at the other end—I sign up a new customer and go through the process of getting the certificate, what do I say five years from now when I do not have enough to serve the customers I had before I added the new one? I have a serious question whether what I have done today because I had some excess deliverability was in their best interests. I should have left the new customer alone and kept my deliverability and choked a little bit on take-or-pay so that they would have gas six and seven years from now that I may not have. That is the problem I am referring to. When we have limited supply, which we are going to have in my judgment over time, it puts a tremendous pressure on pipelines (1) not to take on new customers when they may not be able to serve all their customers sometime in the future and (2) to continue to buy, to serve those customers in the future, which creates problems today.

GORDON GOOCH: There is a big risk in this field of trying to treat the symptoms rather than the cause of the problem. People identify certain problems and say we need some regulatory fix; we do not look at the underlying cause and try to address it.

Consider, for example, this prohibition strategy—just outlaw take-or-pay, just outlaw most-favored-nation or any sort of indefinite price escalator. The question then becomes, What in the contract sets the price? Is there something in the contract then that you look at to set the price? Maybe it is an old contract, and the long-fixed price is twenty-five cents or thirty cents or fifty cents. Does that then become the contract price because you have outlawed the pricing provisions that the parties agreed to apply in a decontrolled scenario? Is that fair? I get to keep your gas; I don't have to pay you for it anymore; and if you didn't anticipate that I was going to get your contract outlawed and didn't protect yourself, you're on your own. As long as we have the allocation controls that say even if the contract has expired, I cannot sell my gas somewhere else, I will have to continue to sell it to you until I go to the federal government and get a piece of paper that says I can sell to somebody else or quit selling to you. It is a wonderful deal. You outlaw all the high contract provisions, and you get to keep the gas. From a certain perspective that is a very attractive alternative.

Although the INGAA proposal does not have that flaw, it, nevertheless, in my view is a regulation scheme because the abandonment requirements, the right-of-first-refusal requirements, are maintained so that in the event of a dispute in a contract still covered by section 104 that cannot be reconciled by the parties, you still continue to sell the gas. There is no reciprocity in any sort of bargaining.

There are significant differences between 107 gas and what will be deregulated 102; the main difference is that, for whatever reason, the Congress made it very easy to bid for the 107 gas. So there is a small amount of deregulated gas against a deeper cushion; even if it is at the 102 level, it is still a cushion. The 107 gas is prevented by law from triggering an indefinite price clause in other categories of gas. It is a little easier to buy it that way because there is no risk of having 102 and 103 gas go up to that price. Finally, there is such a high incremental pricing threshold on 107 gas—I think 130 percent of oil. No wonder people can pay $8 or $9 or $10 for the 107 gas; there is not even a risk that the pipeline will have to price any of it, or very much of it, incrementally, even if there is a return to rule 1 and an adjustment of the price caps back to the number two oil price. So the 107 situation really presents a different set of problems that you can reasonably anticipate happening under an NGPA as is.

MARK COOPER: I have the distinct feeling of walking in at the end of a very long poker game, in which everyone has broken even and they are trying to talk one another into starting all over again. A lot of negotiations have gone on, and proposals have been there and back and forth. Not having been in on the whole poker game, I want to ask a question and see what kind of responses I get to it. What would happen if the Federal Energy Regulatory Commission were to find—as everyone here agrees—that it is impossible to sell gas at $8 or $9, it violates the marketing-clearing price? Therefore, when someone contracts for gas at that price, what they are really doing is selling the new gas that they contracted for at the market-clearing price and selling old gas at the market-clearing price by averaging, which means that in essence they have violated the ceiling on old gas prices—that is, they have cross-subsidized. Wouldn't that have a disordering effect on bidding practices? If it were coupled with a very strong title II rule that revealed irresponsible bidding behavior directly to price-sensitive consumers—that is, if there were a rule that found the effective alternative price and major industrial users paid that alternative price through an incremental pricing law—wouldn't that create a distinct quantity of order in the gas markets? All of that seems to be well within the legal powers of the FERC.

MR. GOOCH: First, I think your premise is incorrect; it seems to me extremely strained to say that under the Natural Gas Policy Act as written the FERC has the power to put ceiling prices on natural gas. I know others are going to be litigating this, so there is no point in debating it, but I think that is your first fallacious premise. The second fallacious premise is that the FERC in its infinite wisdom can set a market-clearing price and hold everybody to it. We have been through that enough; there are not many people willing to go by that.

You say incremental pricing will solve this. Let's take a half a step back and look at that. If you review the statements of Mr. Dingle and the economic projections that came out on the House side at the time the NGPA was passed, you will see that Mr. Dingle and his colleagues did anticipate a fly-up problem after 1985, did anticipate the risk that holding some gas prices down might lead to overbidding. Incremental pricing was placed in the Natural Gas Policy Act for two purposes, one of which was to provide a market-ordering device. We can debate whether the way it was done would work or not. My conclusion is that it could work only partially. Your point that I think very important is that if we leave the NGPA as it is and frustrate every effort to get at the cause of the problems, the FERC will have little option but to resurrect incremental pricing as a market-ordering device after 1985.

MICHAEL CANES: A quick comment and then a longer one. The quick comment is that I found very interesting Robert Leone's analysis that there might be multiple ways in which the rents would be dissipated. I also found interesting Benjamin Schlesinger's agreement.

The longer comment has to do with something Catherine Abbott said, which is that one of the problems with regulation in the market is that it ends up begetting more regulation. It rang a bell in my memory: we just got out of a long period of regulation of oil markets, in which regulations begat regulations over a period of years; indeed, many of the kinds of objections that were raised against oil decontrol are being raised today. There were some who argued, for example, that there would be a big price spike if we deregulated. There were some who said there would be no price spike, that prices would be independent of whether we decontrolled or not. Products prices at least would be independent of whether we decontrolled or not. Economists in the industry and economists outside the industry generally said there would be price effects; that these would have consumption and production effects; that these would have an effect on imported oil; that this might have effects on the OPEC price, on the value of the dollar, on the cost of other imported goods; that while consumers in oil markets might pay more, consumers in general would find themselves better off. Many of the same phenomena appear to exist in the natural gas market.

There was no contracts problem as such in oil decontrol, but there was something called the pull-outs problem. It was alleged that if we simply decontrolled, suppliers would pull out of some areas, regions of the country would find themselves without adequate supply, shortages would exist, and so on. We did decontrol, and those predictions did not come to pass. While the distribution network or the sequence by which crude oil is ultimately transferred to the consumer is somewhat different in oil markets from gas markets, there was concern about certain processors and distributors, many of whom called for the retention of controls, said that the discipline of the market would not prove beneficial to the consumer. The discipline of the market not only has proved beneficial to the consumer but also has proved beneficial to many of the downstream elements that thought it would not.

MR. ZYCHER: Let me repeat a couple of the points that Mr. Gooch made. If we start passing rules to the effect that we cannot clear the market under an NGPA, then the decontrol provisions in NGPA go out the window and we are back with the *Phillips* decision once again. Furthermore, the incremental pricing, the resort to that sort of allocation scheme, puzzles me just a bit since we cannot force people to use gas. It seems to me not very likely that incremental pricing will have much of an effect under any state of the world in the long run, at least if people can move out of gas into something else. It is difficult for me to conceive of a world in which industrial consumers can be forced time after time to continue using gas at $9 or whatever the price may be.

MR. GRENIER: I wanted to respond on the incremental pricing point, too. If we look at the statute and at the inconsistent goals that were sought in the incremental pricing provisions, we realize that there is no way it can really work. Let's assume the worst from the industrial point of view, that the cap is restored to the price of number two oil, a goal that I think the Consumer Energy Council might well support. Then we could probably begin to see some measurable load loss on a lot of systems, and the statute has a built-in mechanism to correct that—it tells the commission to reduce the price to that of number six oil. Some other kind of incremental pricing might not have that limitation and might act as a market-ordering device, but we cannot force people to pay uneconomic prices. We just cannot legislate that.

MR. COOPER: First, it is interesting that the legality of reaching a cross-subsidization rule is dismissed out of hand when the legality of legislating all

117

kinds of contract clauses or decontrolling old gas is never questioned. I think what is legal and illegal, what the FERC can do or not do, is in significant measure what it wants to do and is willing to fight in the courts. I think they could proceed with that kind of rule outlining cross-subsidization and pursue it aggressively just as easily as they can do a lot of other things.

Second, obviously I recognize that incremental pricing is one of the most detested pieces of legislation the Congress passed, certainly in the 1970s, but it is based on a premise that I would like to see tested. We have heard constant testimony that some form of number six oil market is where the market-clearing price is set, and that is exactly what incremental pricing says. We can find a market in which it is supposed to clear, and that is the tail end of the market that drives the whole market. It is interesting that everyone is thoroughly convinced that incremental pricing cannot possibly work, when in fact it stimulates the market. I would be willing to accept something other than the price of number two oil. I would be willing to accept an effective alternative fuel price because that is what we are talking about. We started with the assumption that the effective alternative fuel price was the boiler market. It seems to me logically inconsistent to assure me that the market clears in the industrial boiler market, which is essentially number six oil, but cannot clear in that market under an incremental pricing rule, which is essentially number six oil. It is just not consistent; it is a question of willingness to test the proposition.

ROBERT MEANS: On the question of section 107 gas as cross-subsidization, probably one would at least have to distinguish between those cases in which a package is offered to a pipeline containing both 102 and 107 gas from the case in which a pipeline is purchasing 102 from one producer and 107 from another. It has the alternative of purchasing or not purchasing the 107, but if it wants more supply, that is the market with which it is confronted.

On the incremental pricing, Catherine Abbott said we were heavily involved in the simulation of its effects, but the basic problem is that it does not simulate the market. I am not sure this is a retrospective economic loss or to what extent it was in the minds of the Congress or the congressional staff. It looks like something that set out to apply the inverse-elasticity rule as a second-best solution, where there is a revenue constraint so that not everyone can be charged the market-clearing price. Then it stumbled over (1) the success of the lobby of the electric utilities that exempted them, which means that it is no longer clear on which side there is a greater price elasticity, those that get the higher price or those that get the lower price as a result of incremental pricing; and (2) the representatives of the consumer interests, who wanted to make sure that the incremental pricing stopped at precisely the point where it would start to work from the standpoint of inverse elasticity. So it suffers from internal contradictions. About all we prove is the internal contradictions, not whether one could in fact devise a workable inverse-elasticity rule.

MR. ZYCHER: Mr. Cooper's comments have intrigued me. It seems to me, Mr. Cooper, that you are saying you don't like cross subsidies when they go in one direction but are not too worried about them or about incremental pricing when they go in the other direction. Is it really in your view an appropriate goal of public policy to make sure that cross subsidies go in one direction that some group listened to by Congress happens to favor and not in the other direction? Let's be clear here that cross subsidies are not really at issue; it is in which direction the cross subsidies go.

MR. COOPER: Two things—one, that is the equity statement in the NGPA, which is a normative statement; and two, irrespective of the fineness with which it operates, it is intended as a market-ordering mechanism; depending on how you write the rule, it provides more or less market order. The rules that have existed would not provide much market ordering. One can conceive of a better set of rules, but I readily admit that the NGPA entails a certain equity decision, as the Wright paper and others have said, and that is a normative statement. The NGPA sets the specific kinds of ceilings and makes decisions about the distribution; we can proceed with that or not, and it has not been pursued.

MR. ZYCHER: That phrase "market ordering," is a bit ambiguous. Let us use a phrase that I think captures the essence of what we are saying but that is much less ambiguous, the "wealth transfer." What you seem to be calling a market-ordering problem, Mr. Cooper, is what I would call a wealth transfer problem or, to put it differently, there seems to be a view among some people that price controls help the poor. That is not obvious for at least four reasons.

First, to the extent that the wealthy are precluded by price controls from consuming some good, X, they will instead consume goods Y and Z. It is no longer clear that, given the totality of all goods, the poor end up better off because of the price controls. Second, it is not enough to say that we use curtailments or some allocation mechanism other than price rationing to conclude that the poor are better off. You have to specify what the rationing mechanism is and then show that the poor really are better off under that alternative rationing mechanism. Third, price controls are analytically equivalent to a tax on producers or on owners of productive factors. To the extent that some productive factors, especially labor, are owned by the poor, it is not at all clear that price controls make the poor better off. Fourth, I think we all agree the NGPA makes society poorer as a whole. Reductions in aggregate wealth will be borne, at least in part, by the poor.

The market-ordering discussion tends to cloud what is really going on; there is a fight over wealth, and there is an implicit equity argument being made by some people to the effect that price controls make the poor better off. That argument has not been demonstrated to my mind.

MR. GRENIER: Without meaning to prolong this incremental pricing debate, I would just point out that, as Means and others have noted, by the time Congress had finished with the incremental pricing provisions, a lot of politically sensitive groups were exempted—agriculture, the electric utilities, and others. The only kind of incremental pricing that could work would be something that would have an immediate political response; if we really wanted a market-ordering device, we would incrementally price residential loads, and all the state commissions would immediately become quite sensitive to high prices. That is the only one that would really work and the only one, of course, that would be impossible for any Congress to enact. Incremental pricing ought to be laid to rest. Unfortunately, the courts are not going to let it be laid to rest for a while, but it really ought to be.

MR. EARNEST: I have one comment on the renegotiation issue. I should have made it earlier. I think Catherine Abbott quite properly points out that this is a behavioral question. I am reminded of Pavlov's dog and the fact that you either shock him or entice him. The INGAA proposal, by setting up the capping mechanism, would create some incentive for renegotiation, which is what we all say should happen. To the degree that it would, I think it may be shock or may be enticement; I don't know which.

MS. ABBOTT: I think Robert Leone made a very good point in cautioning us not to set ourselves up so that if we do not find a particular correlation in the behavior of pipelines, we somehow destroy the arguments against partial decontrol strategy. I do think it important to identify real world misallocations to the extent that we can, because there are people who are not convinced primarily by ideological arguments that it is a bad thing to continue price controls on old gas. In other words, I think there is an effort being made here to be fairly specific about exactly what the misallocations are and where they are occurring and to make the point that they may be big and that the problems associated with continuing price controls on old gas may be big. In defense of the effort, it is unidimensional because it is the easiest thing to identify. The problem in trying to point out where cushions may be dissipated in a regulated industry is that we do not have anything to compare it with. We do not have gas pipelines that are not in essence regulated.

On the contract points, I do think that the INGAA proposal is a substantial advance over the prohibition strategy. That is, I do not understand advocates who can argue yes, we can decontrol certain categories of gas or old gas eventually because the wellhead market is workably competitive, but somehow we cannot trust the contract terms that parties in that market are willing to negotiate; so we should restrict certain kinds of contract clauses. In addition, it is very difficult historically to argue that indefinite price escalator clauses per se are the product of regulation since in the intrastate market in 1978 roughly half the contracts were covered by such clauses. I am not saying they are good; I am simply saying that we do observe them in the one gas market that we have had that was not regulated. It may well be that other contract forms will turn out to be more attractive to pipelines and producers, but prejudging what those contract terms ought to look like seems to me to be tying the gas market up too much.

Part Three

The Gordian Knot
of Natural Gas Prices

Henry D. Jacoby and Arthur W. Wright

Federal policy toward natural gas prices is once again the subject of national debate. Thought to be settled once and for all by the Natural Gas Policy Act of 1978 (NGPA), it reemerged as an issue in 1981. The proximate causes of the renewed controversy include Ronald Reagan's campaign promise to seek well-head price decontrol and the administration's attempts (until March 1982) to find a workable decontrol proposal. But the wellsprings of the problem go deeper, to the history of gas price regulation, to changes in energy markets since 1978, and to serious defects in the NGPA itself.

The problem of gas policy has grown increasingly complex. One is reminded of the Gordian knot of ancient myth. According to an oracle, whoever could untie the knot would rule over all Asia. Many tried and failed, until Alexander the Great—unable to untie it but determined to fulfill the prophecy—cut the knot with his sword and went on to conquer as much of Asia as the ancients knew.

In this paper we examine the main policy alternatives, as we see them, for dealing with the tangled knot that gas price policy has become. The question we address is, Given the current situation in gas markets, what might we consider doing, for what reasons, and with what consequences? In seeking an answer, we first need to understand how we got into such a mess in the first place. The next section of this paper reviews the history of federal controls on natural gas, along with the version of resource economics that seems to have

This paper draws upon research sponsored by the Massachusetts Institute of Technology Center for Energy Policy Research. Any opinions expressed are those of the authors alone. Thanks are due to Paul Carpenter for valuable research assistance and collegial advice. We also wish to thank Michael Canes of the American Petroleum Institute, Mark Cooper of the Consumer Energy Council of America, and other participants at the AEI conference on natural gas deregulation for helpful comments and criticisms. A slightly different version of this paper appeared in *The Energy Journal*, October 1982, pp. 1-25.

provided the rationale for their imposition. The paper then turns to an analysis of broad options for dealing with the gas price problem. The final section summarizes the analysis and discusses the pros and cons of the options considered.

To anticipate, the history of federal policy on natural gas consists of a long series of attempts to effect an equitable distribution of its benefits between consumers and producers and among regions of the country. Each attempt has run up against the facts of resource economics and changing conditions in energy markets and has eventually come to be viewed as intolerably disruptive and wasteful. Nevertheless, while efficiency losses may have provided the impetus to change the policy, distributional issues have dominated the design of each new set of controls—thus starting the next round in the cycle.

We consider four policy options: two extreme measures and two that lie in between. At one extreme is the immediate and complete deregulation of field prices—the policy equivalent of taking the oracle literally and trying to untie the knot. Not even in myth, however, was such a neat, efficient solution possible; quick deregulation serves more as an economist's benchmark than as a live option. At the other extreme is sticking with the current policy, which will give us partial decontrol in 1985 or 1987, provided the NGPA is permitted to run its course. This option—the policy equivalent of ignoring the oracle and hoping to conquer Asia anyway—is a live one but runs several unattractive risks. Of the two intermediate options, one is to amend the NGPA to provide for complete but phased decontrol, as proposed in mid-1981 by (among others) Representative Philip Gramm (Democrat, Texas) and the White House Cabinet Council on Natural Resources and Environment. This option, which is similar to the first except for the gradual phase-in, can be likened to Alexander's way of getting around the oracle; unfortunately, it requires a leader with Alexander's resolve and resourcefulness. The other intermediate option is to unify regulated-gas price ceilings by raising those on "old" gas closer to those on "new" gas and then to let the NGPA play out, as suggested early in 1982 by C. Michael Butler, chairman of the Federal Energy Regulatory Commission (FERC). This option would reduce the price "fly-up" trauma expected in 1985, yet perhaps avoid a full-blown renegotiation of the politically charged equity settlement. It thus looks like the policy equivalent of loosening the Gordian knot a little, in hopes that it will unravel by itself in 1985.

Origins of the Current Policy Tangle

The Beginnings of Field Price Regulation. Most of the current problems with natural gas prices at the wellhead can be laid to the NGPA. That piece of legislation did not, though, simply appear one day on President Carter's desk. Its roots go back over forty years to the Natural Gas Act (NGA) of 1938. It will be helpful in understanding current issues of gas price policy to sketch the progression of federal policy on wellhead gas prices since the late 1930s.[1]

The NGA was directed primarily at the interstate pipeline segment of the gas industry. There were efficiency concerns about, for example, backward vertical integration by pipelines into gas extraction and distributive concerns about retail price burdens and regional access by users to the new and relatively cheap fuel. Under the act, the Federal Power Commission (FPC) set up procedures for federal regulation of pipelines carrying gas in interstate commerce and, most important, for federal control of the prices charged to distribution utilities and other customers given the prices paid for gas at the wellhead. The NGA expressly did not cover wellhead prices, although before its enactment the possible regulation of wellhead prices was debated on grounds of both efficiency and equity.

The extension of federal regulation to prices at the wellhead dates from a Supreme Court decision in 1954. As the industry expanded after World War II, state and regional interests struggled over the distribution of the benefits. Producing states imposed export taxes and controls on gas in an attempt to increase both their shares of gas revenues and their own access to gas supplies. Consumer states fought back, and one of the ensuing court cases—*Phillips Petroleum Co.* v. *State of Wisconsin*—became the vehicle for the Supreme Court decision that conferred on the FPC the responsibility for setting field prices of gas sold to interstate pipelines. We note here, for later reference, that because NGA covered only interstate gas sales, gas that stayed within state borders was left unregulated. Eventually this distinction spawned a set of intrastate wellhead markets not subject to federal price controls.

The FPC met its new price-setting responsibility with three different regulatory regimes, in three different periods. Until 1960 it attempted to apply rate-of-return regulation, case by case, but that soon proved unworkable. From 1960 to 1975, it set "area" rates based on rate-of-return principles applied to broad classes of similarly situated producers. Finally, from 1975 to 1978 (when the NGPA was passed and the FPC gave way to the FERC) the FPC set much higher "national" rates that were only very loosely tied to rate-of-return calculations.

The Flawed Basis of FPC Regulation. The economic rationale that underlay FPC wellhead price policy during the first two of these regulatory regimes rested on a fixed-stock, or Old-Mother-Hubbard's-cupboard, view of natural gas supply. In this view, the long-run supply curve of gas was essentially flat over the range of production expected in the medium term but would then become very steep.[2] In effect, for a while the available in-ground resources of gas could be found and extracted at near-constant marginal cost, implying near-constant wellhead prices even though demand was growing. Eventually, however, and in the not too distant future, the rate of gas output could not be expanded regardless of how high the price went. Thereafter, increases in

127

FIGURE 1
The Rationale of Federal Wellhead Price Policy

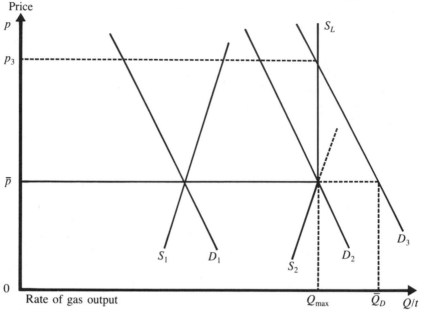

demand would simply raise wellhead prices without eliciting further increases in output. When Old Mother Hubbard went to look for new gas reserves, she would find the cupboard bare.

This view of natural gas supply is depicted in figure 1, which represents a time-lapse exposure of the comparative-statics adjustment of the industry. S_L is the postulated long-run supply curve of natural gas—horizontal up to Q_{max}, at which point it becomes vertical. S_1 and D_1 are representative fully adjusted short-run supply and demand curves, respectively, in the medium term. They intersect at $p_1 = \bar{p}$, which is equal to the presumed constant long-run marginal cost of gas (at the wellhead) throughout the medium term. S_2 and D_2, in a later period, are the last such short-run curves before S_L becomes vertical. When demand increases to, say, D_3, the market price will rise to p_3, but gas output will remain constant at Q_{max}.

A wellhead price of p_3 would generate a pure economic rent equal to $(p_3 - \bar{p})Q_{max}$. That is, given D_3, the gas would be worth $p_3 Q_{max}$, but it would cost only $\bar{p}Q_{max}$ to produce. As with any pure rent, a distributive issue would arise over who gets the rent—producers (for no extra productive effort), or users of gas, or a third party such as the government. A related issue is whether the rent should be collected and distributed through the market (giving it to producers), through legal price ceilings (giving it to users), or through a formal tax-transfer mechanism (allocating the rent though the government budget).

128

Price ceilings are, of course, an informal tax-transfer mechanism that bypasses the budget. Producers and royalty owners are taxed, and consumers receive the proceeds in the form of lower prices. Price ceilings do, however, eventually create excess demand (for example, $Q_D - Q_{max}$ in figure 1), and so they often require a system of allocation—a form of explicit transfer mechanism—to ensure that the benefits actually go to desired classes of consumers. Otherwise, the rent may be misdirected or dissipated—for example, through expenditures to try to gain access to scarce gas.

As noted, after the 1954 *Phillips* decision the FPC chose to collect and disburse the rents informally, by imposing legal price ceilings at the wellhead. In due course shortages appeared, and so the FPC assigned priorities of access to the valuable price-controlled gas. Residential uses received the highest priority and boiler uses the lowest, with public, agricultural, commercial, and industrial claimants arrayed in between. In effect, the economic rent was appropriated through an implicit tax and allocated through an explicit transfer mechanism—outside the Treasury but nonetheless through government intervention on both sides of the wellhead gas market.

From about 1960 through the mid-1970s, the FPC's price-ceiling-cum-allocation mechanism can be said to have proved rather successful "in the small"—for the control of rents on given volumes of gas flowing to the interstate market (mainly gas already under contract when the mechanism was set up). Besides the use of explicit allocations to distribute rents, certain peculiarities of the gas industry—such as the difficulty of reselling price-controlled gas bought from a regulated pipeline—inhibited the kinds of rent-dispersing or -dissipating activities that often accompany price controls.[3]

The mechanism was far less successful, however, "in the large," because the price ceilings had counterproductive side effects both within and outside the interstate gas market. They discouraged investment in new reserves, making the excess demand for gas worse than would be anticipated from a picture like figure 1. The interstate shortage was magnified by the existence of the unregulated intrastate markets, where higher prices attracted gas supplies away from the interstate market. The interregional differences in price and availability of gas distorted interfuel substitution and altered the regional location of gas-intensive industrial activity. The shortage was alleviated through increased oil imports, which of course contributed to other U.S. energy policy problems.[4]

Central to the matter was the defective rationale behind federal wellhead price policy. Neither was the cupboard soon to be bare, nor were new gas reserves available in the interim at constant marginal cost. As is well documented by experience for natural resources generally, the long-run supply of natural gas is an increasing function of price. New reserves were there to be found and developed—for all practical purposes indefinitely—but only at rising marginal cost. But with interstate ceiling prices held more or less con-

129

FIGURE 2
THE REALITY OF FEDERAL WELLHEAD PRICE POLICY

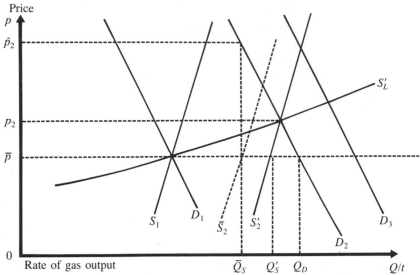

stant in real terms and the gas industry largely convinced that things would stay that way, producers gradually ceased to invest in gas exploration and development for commitment to the interstate market.

The emergence and growth of the interstate shortage of natural gas is illustrated in figure 2. S_1 and D_1, with the equilibrium price of $p_1 = \bar{p}$, are the same as in figure 1. Unlike S_L in figure 1, however, the actual long-run supply, S_L', rises as the rate of output increases. Thus, for demand D_2, the equilibrium price is $p_2 > p_1 = \bar{p}$. The fully adjusted short-run supply of interstate gas given p_2 would be S_2', which would give a shortage at ceiling price $\bar{p}$ of $Q_D - Q_S$. But S_2' will not be realized if the prevailing and expected price is $\bar{p}$; the fully adjusted supply given $\bar{p}$ would be something less—say, $\bar{S}_2$, which would yield an even greater shortage, $Q_D - \bar{Q}_S$.

The shortage will tend to grow as demand increases (say, to D_3) and as the supply adjustments at the price ceiling $\bar{p}$ lag behind. Offsetting the growth in the shortage, of course, will be the dampening effect on increases in demand as decision makers come to expect a mounting shortage.

The preceding analysis gives an accurate, if stylized, picture of what the interstate gas market looked like by the early 1970s. With federal price ceilings held virtually constant, new long-term contracts all but ceased to be written in that market. The only major source of new interstate gas was federal offshore leases, which were regarded for purposes of price regulation as producing interstate gas, whether or not the gas crossed a state line. But the new offshore supplies were not enough to close the gap, and as a result many trunk pipelines

130

and distributors had to institute "curtailments"—reductions in deliveries below the volumes contracted for. Those curtailed first, of course, were the lower-priority users, such as boiler and other industrial customers. Many gas distributors even had to announce moratoriums on new gas hookups for the higher-priority residential and commercial users.

The interstate natural gas shortage of the early and mid-1970s had a number of adverse economic consequences. The federal price ceilings and allocations subsidized residential and commercial heating for those who had hooked up before the moratoriums were imposed. It led industrial customers to switch to other fuels or to migrate to the gas-producing states with their intrastate markets. (The nominal prices of gas were higher there, but they were nevertheless lower than the infinite prices implied by curtailment or denial of a new hookup.) And it created artificial demands for high-priced sources of gas such as imports of liquefied natural gas (LNG), which could be sold by pipelines or distributors by averaging, or rolling in, their high prices with the lower regulated prices paid for other gas.[5]

The interstate gas shortage interacted with developments in the oil market in a way that can be described with little exaggeration as a vicious cycle. Precisely at the time when the gas shortage was beginning to pinch, U.S. oil import policy was changing from artificial restriction to artificial stimulus. The oil import quota, imposed in 1958, effectively came to an end in 1972 (officially in early 1973). The oil price controls that began life as part of the Nixon wage-price freeze of 1971 created excess demands for domestically produced oil. Imports then became the marginal source of oil in the U.S. market, and with no import restrictions (save for a small tariff) the shortage of domestic oil was made up by increased purchases abroad. The interstate gas shortages also increased the U.S. demand for oil, thereby adding to the oil import bill. Then the sharp run-up of world oil prices in 1973–1974 further increased the demand for natural gas, exacerbating the gas shortage.

The drying up of new contracts and the emergence of a serious shortage in the interstate gas market was sometimes interpreted as evidence supporting the fixed-stock view of long-run gas supply. The inference was fallacious, however, because the shortage was created by policy, not by the exhaustion of gas reserves. The shortage in the interstate gas market was accompanied by the steady growth of reserves and output committed to the intrastate markets.

The largest intrastate markets were found in the major gas-producing states—Texas, Louisiana, and Oklahoma. These states had their own gas pipeline systems, necessarily separate from the interstate system. There was intrastate market activity in lesser-producing states as well. One not infrequent form of such activity was the drilling of wells in small fields to provide gas to a local industry.

The intrastate gas markets were the inevitable products of profit maximization under incomplete regulation. With their market-clearing prices, these markets no doubt improved the economic efficiency of the U.S. economy, given the shortage in the interstate market. Yet, the overall allocation of U.S. natural gas production and consumption was wasteful in spite of the intrastate markets. Constrained by state boundaries, wellhead gas prices were unequal across intrastate markets. Further, compared with no price controls or (arguably) even with uniform price controls on all gas, the existence of unregulated intrastate markets led to costly investments (for example, in new pipelines for intrastate sales and in plant relocations) that, from a social perspective, dissipated the gains that were supposed to accrue to users of gas under the interstate price controls. This is not, of course, to fault the participants in the intrastate markets, who were merely responding to the effects of the federal policies that gave rise to the interstate shortage.

The quadrupling of world oil prices in 1973–1974 served as the catalyst for an eventual change in federal policy on natural gas prices—the move to the third of the FPC regimes mentioned earlier. At oil prices of $12 instead of $3 a barrel, gas at $0.20 per thousand cubic feet was seriously underpriced. In 1976, under a new "national" rate policy, the FPC unilaterally raised price ceilings on new interstate contracts. Those ceilings, which reached the range of $1.50 per thousand cubic feet by 1978, were no longer tied to estimated costs of production. Even though it was too little, too late to correct the interstate shortage, the switch to national rates signaled the abandonment by the FPC of rate-of-return pricing and, with it, of the fixed-stock theory of gas supply.

The Natural Gas Policy Act of 1978. The FPC's ad hoc reworking of federal natural gas price policy was seen to be inadequate to remove the shortage in the interstate market. As a result, beginning in 1977 the Carter administration undertook a fundamental redesign of federal gas price policy. The primary goal was to raise gas prices, especially for new production, for the sake of efficiency in both output and use. Presumably out of concern for adjustment costs, the price increases were to be phased in gradually. At the same time, the administration sought to stanch the rent dissipation that was occurring through the unregulated intrastate markets. This was to be accomplished by extending wellhead price ceilings to intrastate gas.

By the time the Carter administration's gas policy initiative emerged from Congress, however, its efficiency goals had been thoroughly intermixed with distributive goals and special-interest favors. The NGPA was a "Byzantine structure" (in one lobbyist's phrase) that achieved some of the administration's purpose but quite a bit else besides—not all of it consonant with economic efficiency. On the positive side, the NGPA raised most price ceilings and scheduled some of them to be phased up to the presumed market-clearing level in 1985. It also achieved a limited measure of market unification between interstate and intrastate gas.

On the negative side, however, the NGPA set up a highly complex set of price categories, replete with compromises, inconsistencies, and perverse incentives. Moreover, the act contained a fatal flaw in its definition of the market-clearing prices to which more than half of all gas was to be raised by 1985. Early versions of the bill tied these prices to world oil prices, however they might change over the intervening years. But as the bill eventually emerged from the conference committee, the target in 1985 was a calculated number, based on the proposition that the oil price in constant dollars would not (or, perhaps more accurately, as a matter of U.S. policy *should* not) rise above the then current price of about $15 per barrel. Without the Iranian revolution and the 1979–1980 price jump, the NGPA scheme could have worked reasonably well in creating a more efficient gas market. But with the oil price increases, the NGPA had in fact laid the basis of what has come to be known as the "market-ordering" problem—in fact, a market *disorder* problem.

NGPA established some twenty categories of gas, according to "vintage" (date of commitment for sale) and other characteristics. Table 1 summarizes the main features of the price categories. Each category had its own price ceiling and escalation formula, and the prices ranged widely—from about $0.25 to several dollars per thousand cubic feet. All "new" gas (from wells spudded after January 1, 1977) was assigned relatively high ceilings (about $2.50 thousand cubic feet), and the escalations were designed to decontrol new gas by 1985 or 1987. The ceilings for old gas were set close to the pre-NGPA regulated prices; they would rise at about the rate of inflation but remain under FERC control indefinitely, even upon rollover (commitment under new contracts when the original ones expire). High-cost gas received relatively high price ceilings, and one subcategory—"deep" gas from wells below 15,000 feet—was deregulated. The price differentials between categories created a built-in incentive for "category creep"—the switching of gas from a lower- to a higher-priced category, often involving rent dissipation, such as the drilling of replacement wells.

The NGPA assigned interim ceilings (involving price rollbacks in some cases) to "old" intrastate gas until 1985, when it too would be decontrolled. These interim ceilings were generally quite a bit higher than those on old interstate gas; thus price unification between interstate and intrastate markets was confined to new gas. The NGPA did not modify the ban on intrastate access to federal offshore gas contained in the Submerged Lands Act and the Outer Continental Shelf Lands Act.

Redistributive goals, with some of the aura of the fixed-stock view of long-run gas supply about them, were readily apparent in the NGPA and in an accompanying bill, the Power Plant and Industrial Fuel Use Act, commonly referred to as the Fuel Use Act (FUA) of 1978. Title 2 of the NGPA provided for "incremental pricing" (not to be confused with marginal-cost pricing), designed to load the burden of new, higher-priced gas onto industrial users of gas (if they were available to bear the burden) rather than residential or

133

TABLE 1
NGPA NATURAL GAS CATEGORIES

Section and Description	1980 Production Estimate (Tcf)[a]	Average Price 3/81 ($ per Mcf)	Date of NGPA Deregulation	Comments
102 New natural gas	2.2	2.73	1/1/85	Includes outer continental shelf
103 New onshore production wells	2.5	2.47	1/1/85, 1/1/87	1/1/85 for wells deeper than 5,000 feet
104 Old interstate gas	7.8	0.25–1.99 (average ≃ 1.25)	Not deregulated	Price escalates at monthly inflation adjustment
105 Old intrastate gas	5.3	0.50 – ? (average ≃ 2.00)	1/1/85	
106 Sales under "rollover" contracts	0.9			
Interstate		0.75	Not deregulated	
Intrastate		1.37	1/1/85	
107 High-cost gas	0.4	Market (≃ 7.00)	11/1/79	107(c), "tight sands" gas is not deregulated
108 Stripper well gas	0.4	2.92	Not deregulated	<60 Mcf per day
109 Prudhoe Bay and other gas	Negligible	1.99	Not deregulated	
Total	— 19 Tcf			

NOTE: Tcf = trillion cubic feet; Mcf = thousand cubic feet.
a. Based on figures from U.S. DOE, Office of Policy, Planning, and Analysis. There is substantial disagreement over these estimates, even among DOE sources; no single estimate of these quantities is widely accepted.

commercial customers. The FUA contained "off-gas" provisions that— out of a concern not to waste precious gas on low-value uses *and* in recognition of shortages that would remain under the NGPA because of continued underpricing—would curb access to gas by boiler users in general and electric utilities in particular. The FUA can be viewed as a more complex, statutory counterpart of the earlier shortage-management allocation regulations, just as the NGPA was the more complex, statutory counterpart of the earlier price control regulations.

Mention has already been made of the low price ceilings the NGPA maintained on old interstate gas. Those ceilings on more than a third of all currently flowing gas, together with incremental pricing, provide what is called a cushion for residential, commercial, agricultural, and small industrial users. They buy gas from distributors, who in turn buy from interstate pipelines, who in their turn can roll in more expensive gas with the cheap old gas and still keep the average price low enough to avoid loss of market. It is this gas cushion, coupled with the existence of some unregulated sources of gas, that is at the core of the current market-ordering problem.

Market disorder in the gas industry takes several forms. First, rolling in low-priced regulated gas with other sources results in the underpricing of gas to large numbers of users. Second, the price ceilings that create the cushion provide an artificial disincentive to drill for categories of gas that are regulated. Third, the cushion supports the bidding up to absurd levels (well above most estimates of the decontrolled market price) of gas in the one deregulated category—"deep" gas— creating an artificial incentive to drill deep (and high-cost) wells. Fourth, the cushion provides some pipelines with an advantage over others in obtaining new gas reserves: those pipelines with relatively large cushions (that is, large amounts of cheaper regulated gas reserves) can use roll-in pricing to outbid pipelines with only small cushions. The most dramatic effect is the advantage given interstate pipelines (which have large cushions of price-controlled old gas) over intrastates (which have much smaller cushions), but the interstates themselves also differ in their abilities to bid for new reserves.[6] The result of the dispersion in pipelines' cushions is a crazy-quilt pattern, across pipeline systems and among regions, of average prices of gas delivered to large industrial customers and at "city gates" to distribution companies.

Options for Restoring Order in Natural Gas Markets

We suspect there would be little disagreement, across a wide spectrum of participants in and observers of the gas industry, with our characterization of the current natural gas policy problem and its history.[7] That could not be said, however, of views on what we should do to deal with the problem. In this section, we compare and evaluate the principal options for wellhead price policy. Our primary focus is on their relative effects on efficiency, particularly how they cope with the market-ordering problem. But we also attempt to factor in the equity issues that accompany all the principal alternatives. As noted earlier, we group the many possibilities under four broad headings:[8]

1. Immediately decontrol all wellhead prices.
2. Do nothing, either legislatively or administratively.
3. Amend or replace the NGPA legislatively, to decontrol "new-new" gas at once and to phase up the prices of all currently regulated gas to market-clearing levels over three or four years.
4. Raise and "unify" the prices of old gas—otherwise do nothing.

As further background for the evaluation of these options, we need to discuss the principal arenas of conflict, who the main combatants are, and certain imponderables that will affect or even determine the outcome quite apart from the efficiency and equity concerns that we adduce. Then we examine several issues that are part of the larger gas debate but are largely separable from the choice among the four major policy options as we define them.

Arenas of Conflict, Combatants, and Some Imponderables. There are two principal arenas of conflict in natural gas price policy. One, of course, is the Congress (together with the administration). Our options (1) and (2) would involve amendment or replacement of NGPA and perhaps other legislative changes, and option (4) could do so too. The other main arena is the Federal Energy Regulatory Commission (FERC), an independent agency (in but not of DOE) charged with implementing the NGPA. Under the provisions of the NGPA, FERC has already proposed changing the price ceilings for certain kinds of "incentive" gas (such as "near-deep" and "deep-water"). In addition, FERC is exploring the possibility, under existing law, of making major revisions in the prices of old gas (called, in the vernacular, section 104 and section 106 gas—see table 1).

The combatants represent a mixture of competing interests that is exquisite in its complexity. There is the natural and traditional conflict between consumers—residential, commercial, agricultural, and small industrial users who receive an economic rent on price-controlled gas—and gas producers and royalty owners. This division also pits region against region, with the gas-producing southwestern states against the gas-using eastern and far western states. Furthermore, there is conflict between users of gas. Large industrial customers tend to support deregulation in principle; however, they have differing views of the proper pace, depending on the weight of gas cost in the firm's overall cost of production, on the relative ease of substituting other fuels, and on the firm's relative position on the priority list for curtailment.

The distribution companies are anything but unified. As a group, distributors see only trouble in wellhead price decontrol. But some advocate rapid deregulation of all but old gas, some are happy with the NGPA as it is, and

136

others would prefer to extend controls beyond 1985, perhaps more than the NGPA now provides for.

Then there are the natural gas pipelines. One must first of all distinguish between the interstate and the intrastate pipelines. The intrastates feel disadvantaged, compared with the interstates, on new gas (because of the ban on intrastates' access to federal offshore gas) and on deregulated or section 107 gas (by the interstates' bidding cushion). Understandably, the intrastates are more in favor of substantial changes in existing policies, including decontrol, than the interstate pipelines. Even the interstates are generally in favor of ending wellhead price regulation. But the differences in their circumstances are so great that there is a wide diversity of views on the details of decontrol, the implications for existing contracts (on which, more presently), and so on.

Even producers of natural gas are not wholly unified on deregulation. Those with large quantities of deregulated gas (section 107)—many of whom have specialized in deep drilling and production since passage of the NGPA— do not want any policy changes that would wipe out the old-gas cushion, for the cushion provides their market by enabling the interstate pipelines to bid prices of $10 or $11 per thousand cubic feet for their product. Producers with large quantities of old gas, who thus provide the cushion, naturally feel otherwise.

And on, and on. Small wonder that veterans of the gas industry wars liken the process of negotiating the NGPA to laying out a path through a minefield. Moreover, it is unlikely that the NGPA can be amended or replaced without involving and somehow reconciling the competing interests, now more fractious than before because of the new special interests it has created. To ignore some or all of them would be to blunder blindly through the minefield, letting bodies fall where they may. In whichever arena the action occurs, the struggle will be fierce, involving pressures beyond our capacity as economists to comprehend. Nevertheless, we have to admit that judgments about likely political outcomes do color our discussions of the policy options.

There is also a silent participant in the process that should not be forgotten. That is the world oil cartel, the Organization of Petroleum Exporting Countries (OPEC). The NGPA, for all its shortcomings, still would have worked reasonably well if only OPEC had gone along with it. In 1978 as now, a solution that looks good with stable or falling world oil prices may look terrible after another oil price shock. Or a proposal keyed to rising oil prices may in fact prove a waste of effort if oil prices continue soft through 1985. We see no way to build the imponderable of OPEC into the analysis other than to ensure that any policy proposals are tested against a wide range of future oil prices.

Uncertainties about supply and demand should also influence our evaluation of different policy options. Little is known about the supply response to rising gas prices, in no small part because the data are so polluted by the prolonged period of regulation-induced shortage and the several shifts in federal regulatory policy. We also know little about what happens to investment in gas and the time pattern of its extraction when there is an expectation of a gradual rise or of a sharp jump in prices within a few years. Even the size of the price jump is uncertain, because it depends on what happens to price ceilings under the NGPA, on decline rates in the various price categories—and on the silent participant. Moreover, there is uncertainty about the magnitude of the response of gas users to rising prices, through conservation and substitution. As a result, there are widely varying estimates of the market-clearing price of gas once most categories are freed of price ceilings.

Thus the choice of one or another policy on natural gas prices is a complex gamble. There are significant imponderables about political forces and economic institutions. There also are substantial uncertainties about market conditions, both domestic and worldwide, to 1985 and beyond. The one real certainty is that we *will* choose a policy option, if only our second one—to do nothing.

Some Common Issues. Several issues in the natural gas controversy are common to all the policy options considered and can thus be dealt with before we compare one alternative with another. These issues are incremental pricing (Title 2 of the NGPA) and the FUA off-gas provisions, intrastate pipelines' access to federal offshore gas, and the effect of increases in gas prices on existing long-term contracts.

As noted earlier, the incremental pricing of high-cost gas to industrial users and the off-gas provisions of the FUA were attempts to prevent erosion of the price protection in Title 1 of the NGPA for residential, commercial, and small industrial users. Both measures emerged from a mixture of concerns— for equity (save the cheap gas for residential users) and for perceived problems of efficiency (override the purportedly inefficient fuel choices being made by electric utilities and large industrial users). But whatever the rationale at the time, under the market conditions foreseen in the 1980s, efficiency would dictate repeal of both incremental pricing and the FUA. (The off-gas provision has already been substantially weakened by a provision of the Budget Reconciliation Act of 1981.) To the extent that a given policy change reduced the size of the old-gas cushion, of course, incremental pricing would become a dead letter. To the extent that access to gas ceased to be an issue as gas prices rose, forcing particular users off gas would be inefficient.

Before passage of the NGPA, the restriction on the ability of intrastate pipelines to buy gas from federal offshore leases had the effect of partially alleviating the interstate shortage, which as noted above was aggravated by the siphoning off of gas supplies into the unregulated intrastate markets. The NGPA shut off the siphon on new gas by subjecting it all to the same price ceilings (sections 102 and 103—see table 1). But the ban on intrastate access to federal offshore gas remained in effect. The only defense of keeping it is an equity argument: interstate gas customers are for some reason more deserving than intrastate customers of federal offshore gas. On efficiency grounds, however, the ban makes no sense, whichever of our policy options is adopted.

Perhaps the loudest noise in the debate over natural gas price policy has been raised over the so-called contracts problem. We lack the expertise fully to comprehend all the legal issues, but the economics appear straightforward. The problem arises from a combination of the use of long-term contracts in the wellhead gas market and the sharp shifts in both oil prices and federal gas price policy of the past decade. To simplify, before 1978 the interstate pipelines purchased gas under long-term contracts, on a take-or-pay basis, with escalator clauses that, in the event there was no regulated price, would tie prices to other gas sales in the same vicinity or to the prices of other fuels, such as No. 2 fuel oil.[9] Wellhead decontrol raises the specter of triggering those escalator clauses. The seller's market that motivated the pipelines to sign such contracts in the first place is weakening today, and suppliers are not expected to be so favored in the future—certainly not after wellhead decontrol. Should the clauses be triggered, however, the pipelines that hold them could have to buy large amounts of high-priced gas, but would have declining amounts of low-priced gas against which to roll it in. Many pipelines fear this would price their gas out of the market.

The pipelines would like to get rid of the offending escalator clauses, and so they seek legislative relief—for example, a prohibition on the use of deregulated gas prices to activate the escalator clauses. Producers of gas understandably oppose such measures, arguing sanctity of contracts. Some observers argue that legislation is not necessary, because producers will be willing to renegotiate contracts with the escalator clauses: why should they force the pipelines to buy a lot of gas that cannot be sold? Others quail at the prospect of thousands of contracts being renegotiated or litigated, and they urge a legislative solution. The opinions on the contracts question are so richly diverse, even among people who should know, that only one conclusion is possible: the issue is fraught with uncertainty.

We take no position on the contracts problem, but content ourselves with pointing out that any legislative effort will have to confront the opposing interests head on, as will private renegotiation of contracts. The contract escalators may have smaller or larger effects under the different policy options considered here, but the problem is qualitatively the same in all cases.

The Four Policy Options. We come finally to the four broad policy options that appear to us to span the range of possibilities for the next round of changes in federal natural gas price policy. Right at the outset, we must take up the key issue of the prospects for a "windfall profits tax." Our evaluation is based largely on efficiency grounds, because the presence or absence of a new tax on gas (while it would be motivated by equity concerns) could make a big difference in the effects of the various policies on efficiency. Efficiency gains are, after all, the basis of the case for decontrolling wellhead gas prices.

The four options analyzed, even "doing nothing," all involve rising natural gas prices. If history is any guide, rising gas prices mean there is a chance that Congress will attach some form of windfall tax to any of them. The tax could take a variety of forms, but we assume (from our reading of the trade and popular press) that it would resemble the "crude oil windfall profits tax" (COWPT).[10] In fact, COWPT is not a tax on profits but a differentiated excise tax applied to the difference between prices received and base reference prices. We also assume that a windfall tax becomes more likely, the faster the rate of increase in gas prices relative to the NGPA—or perhaps relative to what was expected under the NGPA before mid-1981, when gas price policy again became a public issue. The total transfer of wealth, from gas users to producers or to the government, will of course be about the same under under all policy options that lead to decontrol *sometime,* except for time discounting and the efficiency losses incurred so long as decontrol is not complete.

A windfall tax is often said to be the political price of getting any increases in wellhead prices above those called for in the NGPA. The term "price" is well chosen, for an excise tax of this type would both reduce short- and long-run supply responses to higher field prices and artificially raise gas prices to consumers. A windfall tax would thus erode the efficiency gains from decontrolling gas prices. It is a classic example of the "leaky bucket" problem, in which income redistribution measures waste resources as well as transfer claims to them from one group to another. For this reason, we argue that a windfall tax of the excise form should be avoided in the interests of economic efficiency. If it is a necessary component of a new policy, then on efficiency grounds one should prefer an excise tax (1) on fewer rather than more categories of gas, (2) on gas under older rather than newer contracts (because of expectations), (3) on gas with a smaller rather than a larger elasticity of supply, and (4) of shorter rather than longer duration.[11]

At the same time, we recognize that the choice of a natural gas price policy inherently affects the distribution of wealth, and significantly so. Hence it is impossible to ignore equity issues in discussing gas policy options. Nor is it prudent to ignore the appeal of a windfall tax in the present federal fiscal climate, with the intense concern over the sizes of prospective deficits. Finally, the prospect of a new gas tax associated with any change in NGPA pricing schedules has led some industry participants (including large-scale producers) to drop their support for decontrol. In short, close attention should be paid to the likelihood that any policy will eventually be accompanied by a windfall tax and to the search for a tax design that will cause the least possible waste given the necessary distributional effect.

We take up the evaluation of the four policy options in the order in which they were presented above—first the two opposite extremes of immediate decontrol and doing nothing and then the intermediate positions of accelerated, phased decontrol and the unification of old-gas prices.

Option 1: immediately decontrol all wellhead prices. Earlier we likened this first option to trying to untie the Gordian knot—something not even Alexander the Great could manage to do. Hence immediate and total wellhead decontrol is more a point of departure for our efficiency-based evaluations than a likely prospect for policy action. Getting wellhead prices to market-clearing levels as soon as possible would permit both demand- and supply-side gains to be realized sooner and more fully than any other option considered.

The arguments for immediate decontrol do, of course, abstract from several complications. First, there would be substantial short-term adjustment costs that would erode the touted efficiency gains. Second, there are possible obstacles to the efficiency gains in the form of continued regulation of gas pipelines and distributors—obstacles under the other policy options as well. Third, a stiff windfall tax would almost certainly be part of any legislation immediately deregulating wellhead gas prices. Indeed, it is plausible that Congress would want to set the tax up according to the NGPA categories and that therefore this option cum tax would have effects similar to those of option 2, "do nothing," except that the government would collect most of the rents (taking the cushion out of private hands) and that old gas would be decontrolled when the tax ended.

Option 2: do nothing, either legislatively or administratively. This policy option is tantamount to hoping it will not matter if we ignore the Gordian knot. It has the disadvantage of perpetuating the present situation with all its inefficiencies and perversities. But the key issue in evaluating it is what would happen to the cushion between now and 1985. And that issue bears in turn on what is likely to happen as we approach the date when prices of large quantities of gas are to be turned loose. Many of the likely outcomes are not good, and in any event "doing nothing for now" only adds uncertainties to an already complex gamble.

The size of the gas cushion will decline between now and 1985 for several reasons. The NGPA itself provides for escalation of price ceilings and for somewhat higher ceilings on rollover contracts, and old gas is steadily being depleted. The cushion is also known to be leaking, through the bidding up of prices on deregulated, section 107 gas and through "category creep"—the conversion of old gas (sections 104 and 105) into new onshore gas (section 103). Precise figures on the magnitude of the components of the decline are lacking. A rough estimate of the total size of the cushion for 1980–1985, including both the scheduled changes and the leaks, is about $30–35 billion a year (in 1980 dollars) for 1982 through 1985 (see appendix). Even if this estimate is high by as much as 25 percent, there will still be enough of a cushion to generate a rather sharp price "spike" on newly deregulated gas in 1985.[12] Such a spike will eliminate the cushion "from above," as it were, by raising the *average* price of gas, which users pay because of the way pipelines and distributors are regulated, to the market-clearing level. Suppliers, however, will receive an array of field prices, ranging below and above the market-clearing price, with obviously deleterious effects on efficiency.

The prospect of a price spike in the early weeks of 1985 raises the question of what the administration or Congress will do as the day of reckoning approaches. Under the NGPA, either the president (by written order) or Congress (by concurrent resolution) could extend the controls on sections 102 and 103 gas for eighteen months. This could be done any time between July 1, 1985, and June 30, 1987. Extending the controls would prolong the life of the cushion, putting off the day of reckoning. *Or* Congress could pass new legislation superseding the NGPA and imposing whatever price ceilings on gas it wanted to. Were this to be done in a hurry—say, early in 1985 after the 1984 presidential and congressional elections—the efficiency consequences could be unfortunate; that it could be done in a hurry, however, is in some doubt given the complexity of the opposing interests that would still have large stakes in the matter. *Or* Congress could let the NGPA expire but then impose a set of excise taxes using the NGPA structure; the undesirable efficiency consequences of such taxes we have already dealt with. *Or* a variety of other things could happen.

Enough said—doing nothing now looks like a thoroughly disagreeable option. It temporizes with respect to the cushion, but then runs the risk of either getting rid of it in an inefficient way, with or without a windfall tax, or not getting rid of it at all because of further temporizing. All in all, therefore, it looks like a bad gamble—worse than necessary.

Option 3: amend or replace the NGPA to decontrol "new-new" gas now and phase up old-gas prices. This option—the policy equivalent of Alexander's putting the sword to the Gordian knot—is what many people since mid-1981 have meant by the decontrol of natural gas wellhead prices. DOE has analyzed it thoroughly, and variants of it have been put forward from several quarters, including the administration, Congress, and interest groups.[13] Excluding provisions dealing with the common issues we covered earlier, all the variants of this option come down to two essential elements: (1) the immediate raising of price ceilings on new-new gas (from wells spudded on a given date, such as January 1, 1983) up to a cap set at (say) 70 percent of the average refiner acquisition cost of crude oil; and (2) the phased decontrol of all existing NGPA categories over three to five years, with the target price set at the level of the price cap for new-new gas and a monthly price escalation formula. In the case of a three-year phase-up, for example, the prices would be allowed to rise by one thirty-sixth of the difference between the cap and the current ceiling.

This option has considerable appeal compared with do-nothing option 2.

The cutting edge of gas supply, new-new gas, gets higher prices right away. The gradual but brief phase-up of NGPA gas prices takes cognizance of short-run adjustment costs but provides the efficiency gains from decontrol within just a few years. (It may also introduce its own adjustment costs, in the form of the withholding of production until prices are higher.) Meanwhile, the cap prevents the cushion that remains during the phase-up from being used to bid the prices of new-new gas above the eventual market-clearing level.

There are three problems with this option. First, rather than curing the cushion problem, it merely treats its symptoms with a price cap—effectively, a price ceiling on "deregulated" gas during the phase-up. If (as seems likely) the price clauses in new gas contracts were tied to the cap, it would build the OPEC imponderable into gas price policy. In the last few months of the price phase-up, any variation in OPEC prices would be translated nearly one for one into the price cap and thus into contractual gas prices. Second, by most accounts the political price of enacting the requisite legislation would be a set of excise taxes on gas along the lines of COWPT. Third, undertaking to write that legislation would be to open up the Pandora's box of complex opposing interests that the NGPA, after eighteen months of struggle, finally succeeded in balancing. To change the metaphor, writing the legislation for this option would mean resurveying the minefield and picking a new path safely through it. Apparently, neither the administration nor Congress wanted to devote the time and effort to doing so in 1982, a Congressional election year, with a tense budget battle to be fought. Note that if this sort of stalemate persists, option 3 will reduce to option 2. Alexander's solution to the challenge of the Gordian knot requires an Alexander.

Option 4: raise and unify the prices of regulated gas—otherwise do nothing. Option 4 we referred to earlier as loosening the Gordian knot and hoping that the natural gas price problem would then unravel by itself come 1985. The idea here is to "decushion" by raising existing ceiling prices to a common higher level, in order to reduce the distortions arising from the gas-price cushion. The worst of the cushion problem comes from the very low ceilings on old (sections 104 and 105) gas, although the price ceilings on (NGPA) new gas contribute to it as well. Two variants of the decushioning option have been put forward—one by William F. Demarest, Jr., the other by the chairman of FERC, C. Michael Butler III.[14]

This option could be accomplished by new legislation. One might argue that, if the bill were confined to old gas, Congress would have less difficulty passing it than a broader bill that reopened the whole NGPA. But one could also argue that, if old gas is the heart of the problem, the congressional battle would be almost as intense as if full decontrol were at issue. Limiting attention to old gas would, of course, limit the reduction in the size of the total gas cushion.

Decushioning could perhaps also be accomplished through administrative action by FERC. Whether FERC has the legal authority to raise price ceilings under the NGPA is beyond our expertise to judge. As to its political feasibility, there is already some evidence. Chairman Butler's trial balloon, to have FERC raise some old-gas price ceilings, evoked a public outcry and the threat of a congressional resolution opposing "backdoor decontrol"; FERC quickly downgraded the matter from a Notice of Proposed Rulemaking to a Notice of Inquiry. Evidently, changing the arena does not mean changing either the conflicting interests or the stakes involved.

This policy option could, of course, have a windfall tax attached to it. Compared with the preceding option, however, the chances might be better of confining it to a subset of categories and perhaps limiting its duration. As we argued above, the more limited the scope of a tax, the fewer the efficiency losses from it. Note that, if FERC did the decushioning, there would be no windfall tax unless Congress acted independently to impose one. This may account for the intensity of congressional opposition and for producer groups' support of FERC action. Thus using the FERC arena could have the effect of forcing Congress, like it or not, to deal with gas price policy.

With or without a tax, decushioning would provide significant benefits. It would reduce the incentives in the NGPA to expend resources for category creep. It would reduce the source of the bidding up of prices for high-cost deep gas. And it would leave less room for a price jump in January 1985 and hence reduce the political pressures to extend the NGPA or to replace it hurriedly with new controls. The more decushioning that was accomplished, the greater these benefits would be—unless there was a threshold beyond which an across-the-board windfall tax would be the price of raising the price ceilings.

The principal drawback of this fourth option is that (like the second one) it would not by itself lead to the decontrol of all natural gas prices. The distortions from the remaining controls (basically, on old gas) would, however, be considerably smaller than under option 2 precisely because option 4 would decushion whereas 2 would not. There may, of course, be something to the argument that leaving any controls in place, no matter how minor their current effects, invites trouble if conditions change (for example, if there were a new oil price shock) and makes it easier for Congress to reimpose broader controls.

Summary and Conclusions

The design of yet another change in gas policy is a complex gamble in the face of great market uncertainties and political imponderables. We can, however, compare the different policy paths open to us in 1982 or 1983 from the perspective of economic efficiency. Admittedly, the ranking that follows reflects to some extent our judgments about what is politically feasible.

Of the four broad options discussed above, the first one—immediate, complete decontrol—would be best. It would give us the efficiency gains on both the demand and the supply sides from decontrolling wellhead prices of natural gas sooner and more completely than any other single path. But it is as impracticable as it is near ideal. It seems we moderns cannot untie our Gordian knot any more than the ancients could theirs.

The worst of the four is clearly option 2, doing nothing. We may already be on this path; sagging oil prices and the "gas bubble" have made it appear to some that we can get through the 1985–1987 NGPA price changes with little trouble. It is a bad gamble, because of the risk that doing nothing now sets the stage for doing a great deal, badly, in a hurry shortly before those price changes. It seems we should not ignore our Gordian knot, either.

There is not much to choose between our options 3 and 4. Both have the same ultimate aim, and both could be successful, though in different ways. Sawing away at the Gordian knot by phasing up all gas prices over three years would result in complete decontrol; but it would require the use of a price cap, and it could well be accompanied by a broad-based, complex, long-lived windfall tax. Loosening the knot a little by price unification would decushion and hence greatly simplify the gas price problem, and we might get away with a relatively simple windfall tax. But it would leave intact the NGPA price control mechanism, which gave us the knot in the first place.

Our list of four broad policy options is a considerable simplification, made for purposes of analysis, of the full range of possibilities facing policy makers. There are other options—mixtures of the "pure" options laid out here—that warrant further study. An example of a mixed gas price strategy would be first to decushion (our fourth option) and then to phase up all gas until markets cleared (our third option). By approaching decontrol in stages, it might be possible to limit both the scope and the duration of new taxes imposed on the gas sector. It might also be possible to avoid the necessity of imposing additional regulation on gas in the form of a price cap to curb overbidding on deregulated gas. A three-part package of decushioning, a three-year phase-up, and a matching three-year windfall tax might be politically viable.[15]

There are doubtless other decontrol strategies, including some that do not fit our elaborate Gordian-knot metaphor. The question of decontrolling wellhead natural gas prices merits our urgent attention. We need to find a workable alternative to doing nothing and merely hoping for the best.

APPENDIX TABLE

SIZE OF THE CUSHION UNDER NGPA ASSUMPTIONS, 1980–1985

(1980 dollars)

	1980	1981	1982	1983	1984	1985
1. Total quantity (trillion cubic feet)	19.36	18.64	17.95	17.15	16.67	17.12
2. Average price ($ thousand cubic feet)	1.78	2.02	2.21	2.36	2.54	4.45
3. Gross revenue ($ billions)	34.5	37.6	39.7	40.5	42.3	76.2
4. Market-clearing price ($ thousand cubic feet)	4.00	4.10	4.20	4.30	4.40	4.45
5. Market-clearing revenue ($ billions)	77.4	76.4	75.4	73.8	73.4	76.2
6. Cushion (line 5 − line 3) ($ billions)	42.9	38.8	35.7	33.3	31.1	0.0

NOTE: Line 5 is calculated without assuming any price effect on quantity demanded or supplied. Partial decontrol, per NGPA, is assumed in the last column.
SOURCES: Lines 1 and 2 from U.S. DOE, Office of Policy, Planning, and Analysis; line 4, our best judgment.

Notes

1. For a good short history of this policy, see Peter R. Merrill, "The Regulation and Deregulation of Natural Gas in the U.S. (1938–1985)," Harvard Energy and Environmental Policy Center, Discussion Paper Series, E-80-13, January 1981.

2. By "long-run supply curve" we mean the locus of long-run, fully adjusted supply-demand equilibria that a competitive industry will trace out as the size of the market expands. Short-run supplies and demands are "fully adjusted" to given market conditions (including but not limited to current and expected future prices) if no decision maker wants to alter his capacity or rate of output.

3. One source of rent dissipation that became increasingly important in the 1970s was the practice known as rolled-in pricing, which is discussed below.

4. Over the years there have been several studies of the efficiency losses due to gas price regulation, and they all indicate the losses are large. In 1977, for example, Robert S. Pindyck made a rough estimate of the net economic gain from phased decontrol, compared with continued FPC area rates ("Prices and Shortages: Evaluating Policy Options for the Natural Gas Industry," in A. Carnesale et al., *Options for U.S. Energy Policy* [San Francisco: Institute for Contemporary Studies, 1977]). His results showed that the net gain, even after compensating losers (users of gas), would be about $700 million in 1978 and rise to over $10 billion per year in 1981 (in 1976 dollars). A 1981 study by the Department of Energy (DOE) compares the playout of the NGPA (including partial deregulation in 1985) with immediate full deregulation in 1982 (Office of Policy, Planning, and Analysis, "A Study of Alternatives to the Natural Gas Policy Act of 1978," DOE/PE/0031, November 1981). This study estimates the present value of efficiency gains from deregulation over the period to 1995 at some $25 billion (in 1980 dollars). While both these studies show that gas deregulation would reduce oil imports, neither takes account of benefits due to the resulting foreign exchange adjustments. Paul Krugman has shown that these benefits could easily outweigh the direct benefits of oil import savings ("Real Exchange Rate Adjustment and the Welfare Effects of Oil Price Control," MIT Energy Laboratory, Studies in Energy and the American Economy, Discussion Paper, MIT-EL 81– 025WP).

5. The rolled-in pricing of high-cost gas of course dissipated part of the economic rent transferred from U.S. producers to consumers. Because lower-cost gas is displaced by higher-cost, there is less pure economic rent left to transfer.

6. Catherine G. Abbott suggests elsewhere in this volume that the intrastate pipelines have a sizable gas cushion, but it is largely on gas committed to large industrial users and hence not available for reducing the average price of gas systemwide. On differences in interstate pipelines' cushions, see Abbott's paper and also that by Robert Means in this volume.

7. An exception would be the school of thought that disagrees with our implicit premise that the wellhead market is workably competitive.

8. A fifth option—repeal the NGPA and keep gas prices low through permanent FERC field-price regulation—we dismiss without detailed discussion. It would amount to "doing something" outside the range of (nothing, everything) mentioned above and would be a throwback to the approach of the 1950s and 1960s.

9. Take-or-pay clauses require the purchaser to pay for the gas whether or not he

takes delivery, so long as the seller is willing to supply contract quantities. They commonly allow for "makeups"—taking deliveries of gas against past payments—over several years.

10. Alternatives include a lump-sum tax on the increased value of reserves and a tax on unusual corporate profits that might be identified with a revision in federal price controls. The lump-sum tax, which has many nice theoretical properties, presents intractable problems of valuation of reserves in the ground and attribution of value to particular policy changes. Likewise, the overwhelming complexity both of the corporations that make up the oil and gas sector and of the markets in which they operate has long frustrated attempts to construct a "profits tax" that could target particular rents for collection.

11. There may be a tendency to equate older contracts with lower price elasticity of supply and vice versa. The relationship will only be valid, however, to the extent that older contracts are correlated with more intensively developed fields. There is probably some positive correlation here, but it is not perfect. This matter bears further research.

12. Earlier we noted the dependence of our entire analysis on world oil prices. Here the effect is on the market-clearing price of gas. The collapse of world oil prices would deflate the cushion, and another oil shock would further inflate it. Implicitly we are assuming that there are no major shifts in world market conditions.

13. Contributors of specific proposals under this option include the Cabinet Council on Natural Resources and Environment, Representative Philip Gramm, Senators Bennett Johnston (Democrat, Louisiana) and Russell Long (Democrat, Louisiana), and the Process Gas Consumers Group.

14. William F. Demarest, Jr., "NGPA Revisited: Implications of the Historical Origins of the Market Order Issue," mimeographed, November 1981; and reports of FERC studies and plans regarding administrative price increases in various trade and popular newspapers, January–March 1982. Demarest proposes decontrolling intrastate prices immediately and using those prices as caps for interstate prices. Under some circumstances, Demarest's proposal is tantamount to full decontrol.

15. A mixed strategy of decushioning and a three-year phase-up was proposed by the authors in a previous paper, "Obvious and Not-So-Obvious Issues in Natural Gas Deregulation," Massachusetts Institute of Technology, November 4, 1981.

Commentary

Michael Canes

I particularly like four things in the paper by Wright and Jacoby. First, there is some mention of the role of producers' anticipations of price controls in generating shortages. That discussion serves as a useful reminder that policy actions can have consequences even before they are implemented. Second, the paper discusses once prominent notions of a fixed natural gas supply requiring government allocation and why those notions have now been discarded. The transformation of thinking about gas is relevant to evolving policy formulation. Third, attention is given to supply elasticity and to the efficiency consequences of any windfall profits tax on gas. Such consequences are clearly of great importance in considering the advisability of such a tax. Finally, the paper identifies relevant issues to be resolved in moving toward a free market for gas. Precise identification of such issues is helpful in formulating ideas to resolve them.

Having said that, I want to raise several points about the paper. First is a matter of clarity. If one does not know what gas market disordering is when one comes to this paper, one is unlikely fully to understand the discussion of that phenomenon. The paper evidently assumes that the reader knows a good deal about market disordering; if that is not so, the discussion is likely to be somewhat confusing.

Second, there is acceptance in the paper of what I call the full price spike. According to this notion, in 1985 the average gas price under the NGPA will exactly equal the average price under full decontrol (see, for example, the appendix to the paper). Most analysts agree that there are tendencies for that to happen eventually, but there is disagreement about how fast it will occur. Some believe that it will occur very rapidly in 1985, perhaps within weeks or months. The paper implicitly assumes it will happen on January 1. But others think that, although the tendencies are there, there are frictions that will keep the NGPA average price below the full-decontrol price for a good long while. These frictions have to do with the behavior of state public utility commissions in granting price increases to distributors and also with how fast pipelines with access to the remaining controlled gas will bid up prices on newly decontrolled gas. Because of this, some think it could take several years for the average gas price under the NGPA to equal that under full decontrol. At a minimum, the

149

authors should state explicitly that they accept the full spike hypothesis as of January 1, 1985, while noting that this is not necessarily a generally accepted view.

A third point is that this paper does not treat OPEC pricing as a matter of policy interest. To be sure, the paper notes that OPEC prices have an important bearing on how the gas market will resolve itself in 1985 under any policy alternative, but OPEC pricing is not itself seen as an object of policy attention. If one is an internationalist, income redistributions beween OPEC countries and other countries brought about by oil price changes are of no consequence. But if one takes the view that it may be possible to redistribute wealth away from OPEC and toward oil-importing countries by putting downward pressure on oil prices, then U.S. gas policy may be a tool whereby such wealth redistribution can be effected.

My next point deals with the practicality of what the authors ultimately recommend. Their preferred policy solution is to dissipate the cushion of price-controlled gas through legislative or regulatory action affecting the old-gas categories. The difficulty with this is that the old-gas cushion is exactly what is causing political conflict over gas price controls in the first place. The representatives from the American Gas Association and from the GHK Companies have indicated that they want to use the old gas cushion to subsidize, in effect, new and exotic supplies of gas. For them that would be an appropriate use of the old-gas cushion. And I suspect that the representative from the Consumer Energy Council will say that the appropriate use of the cushion is to subsidize gas consumers, even though in my view such subsidization would not help consumers generally. In any event, the very heart of the opposition to decontrol of natural gas prices is the desire to maintain controls on old gas; so although the authors' proposal would be sensible if it could be accomplished, it essentially assumes away the nature of the problem.

My final comments have to do with the efficiency and income transfer aspects of natural gas pricing policy. It is not sufficient for economists to say that there are efficiency and income transfer consequences of various policy alternatives. Policy makers do not always understand precisely what is meant by those terms. It is therefore necessary to explain what efficiency gains mean to individuals as consumers or as resource owners and how they might actually manifest themselves with changes in policy. Further, it is necessary to quantify just how great any efficiency gains might be so that policy makers will understand what can be gained by changing the existing law. From a legislator's viewpoint this is important because, if experience with the NGPA is a guide, it is likely to take much time and effort to achieve any such changes.

Regarding the income transfer issue, it is interesting that the participants at this conference have basically accepted the notion that any transfer that might take place between the NGPA and full decontrol is one among producers of gas. From this perspective, the only question is which producers of gas win and which lose. But this is not the common perception in Washington. The common perception is that the transfer would be between producers and consumers. This perception is based on the notion that the average gas price will be lower under the NGPA than it would be under full decontrol. Under these circumstances, it appears from analyzing the gas market alone that consumers will lose and producers gain. But such an analysis is not correct. A complete economic analysis of decontrol alternatives must take place in a multimarket context and explain the nature of *all* the transfers that might take place. Such an analysis reveals, for example, that while consumers in some markets are made worse off under various gas decontrol alternatives, consumers in others are made better off. That is a very important and different finding from one that merely assesses consumer versus producer outcomes in a single market. Similarly, multimarket analysis reveals that producers in some markets gain and producers in others lose under various decontrol alternatives. That too is a very different finding from one that merely asserts that gas producers gain.

To summarize my comments, the authors clearly are in command of the subject matter and understand the issues and policy alternatives. They can contribute more in the future by supplying more quantitative information and by placing gas decontrol alternatives in a multimarket context, which is the only way to gain a correct view of the outcomes of the policy alternatives.

Mark Cooper

I have one difference with Mr. Canes: I think when one does the arithmetic in a general equilibrium situation, one concludes that we should not decontrol. I will accept for the moment the framework, the assumptions, and talk about efficiency gains and losses. I disagree on certain assumptions about the competitiveness of the natural gas market, about the importance of equity and wealth transfers. But I will stick to efficiency analysis.

In looking over the past several years of debate over the effects of decontrol on efficiency, I find two major sources of confusion. First, a number of types of efficiency gains are possible in the energy sector, and they frequently become confused. Sometimes there is a lot of double counting. Since the authors do not do any counting in their paper, they avoid that error. We do have to count, and we should count very carefully. Second, there is a frequent tendency in efficiency analysis to slip between partial equilibrium analysis, what happens in the energy sector, and general equilibrium analysis, what happens in the entire economy. A misconception can easily be created that the partial solution, which is typically positive, means that the general solution will also be positive. In truth, theory is entirely silent on the general equilibrium result. The answer is purely empirical. It turns on the elasticities of supply, demand, and substitution in nonenergy sectors. And available empirical evidence suggests that the general equilibrium solution will be negative

Efficiency analysis will play an important part, and has played an important part, in deciding these issues. When it comes time actually to count the inefficiencies and therefore the potential gains in efficiency that can be claimed for decontrol, they frequently prove to be a lot less compelling than their advance notices, at least for natural gas. Things like curtailment and rationing allocation inefficiences turn out to be small. Efforts to justify natural gas decontrol usually guess at major import premiums, that is, hidden import costs or import price savings, which then are made to carry the ball for decontrol. Those are suspect gains, and the authors recognize the imponderable of what influence we can have in affecting the world oil market.

Another point is that the demand side of the NGPA seems to have worked reasonably well. There have been demand responses to 20 and 30 percent annual price increases. There have not been an awful lot of curtailments. These do not appear to be directly related to the NGPA. There may be some rationing problems in terms of the associated legislation. But on the demand side the potential efficiency gains are not very great.

With respect to the supply-side inefficiencies that have cropped up under the NGPA, there are a number of possible responses. First, a lot more market ordering could occur under the NGPA under a different set of decisions to implement and a different set of rules. Second, of course, there is a great deal of difference of opinion about how bad the disorder is and will be. It may not be nearly as bad as those people supporting decontrol believe, and I will leave the difference of opinion by those involved in the industry to speak to that. Finally, if we do not assume that the wellhead market is workably competitive, the supply-side gains become very small, although consumption gains, demand-side gains, remain.

Let us assume that the market is competitive and that the authors have properly gauged the partial solution. What does that mean for policy? This is the maddening point in the paper. Having analyzed the partial situation in detail, the policy choices are then made in a general economic or political framework. I have no problem with such a framework, but they have not laid the analytic basis for making those decisions. Above all, I take exception to the primary reason that the authors give for rejecting immediate decontrol of all wellhead prices: "There would be substantial short-term adjustment costs that would erode the touted efficiency gains." This appears to be a general equilibrium argument, or at least it should be, because the adjustment costs in the general economy are in fact huge. I do not quarrel with their rejection of immediate decontrol or with the fact that they have chosen to mention, at least in this instance, short-term adjustment costs and efficiency losses. Not at all. My complaint is that thay do not systematically consider and measure those costs in all cases. That is, what are the adjustment costs and efficiency losses of the other options? How long is the short term? Where is their theory of the general equilibrium adjustment to rising energy costs? Policy makers need to know the whole story for each option. In fact, the authors have let a partial equilibrium discussion parade as a general equilibrium conclusion. They have left the field before the battle is even engaged, not to mention won. The simple fact of the matter is that if the elasticity of substitution for gas is low—that is, if it is difficult to substitute for gas in the production of goods and services—then the shock to the economy from rising prices may be very large. The gains in available resources achieved by efficiency improvements in the energy sector may not offset the losses in output and productivity in the nonenergy sectors of the economy. That is where the empirical question comes in. We need to look at the empirical evidence.

Here we might turn to DOE's recent analysis. Let us focus on the single model that predicts a positive economic outcome—the dynamic general equilibrium model. If you look at that model, what you discover is that there are productivity losses in the economy due to decontrol that last for fifteen years. And the losses in productivity or in GNP due to decontrol are larger than the efficiency gains in the energy sector. The model offsets those losses by assuming the absence of a wage-price spiral. It translates all price increases from gas directly into losses in labor income and increases in the income of gas-related capital. There is a transfer of wealth from those with a high propensity to consume to those with a high propensity to save and invest. By raising the aggregate investment rate, the model then predicts GNP increases in the long run. Thus, even in the single model that predicts a positive general equilibrium solution, energy price decontrol policy does not really stand on its own economic merits. In and of itself, it is negative. It appears positive because of the income transfer effect and the increase in the aggregate rate of investment. But one must ask whether even a moderate wage-price spiral built into the model would reverse that.

In conclusion, I agree with the general approach of the authors in attempting to identify efficiency gains and losses. I recognize that if their assumptions are correct, the arguments are correct as well, although I would probably disagree with the assumptions. Even if I accept their assumptions, however, I think that a more careful counting of potential gains in the energy sector will lead to a downward revision of their estimates; at least we will not constantly say how big the efficiency gains will be. Even if I grant their estimates, I believe that the general equilibrium result will be negative. I also think they may have overlooked a perfectly good policy option—making the NGPA work as Congress intended.

Discussion

BENJAMIN ZYCHER: It is difficult for me to believe that an increase in the production of natural gas can make us worse off as a general equilibrium proposition. The problem with the macro models is that they treat decontrol of natural gas just as they treat an increase in international oil prices, perhaps generated by a decrease in Saudi oil production, when in fact those two phenomena are entirely different. An increase in international oil prices makes us poorer, whereas a decontrol of natural gas, by improving the productive efficiency of the economy, is very likely to make us wealthier. Nonetheless, the macro models tend to give us the same sorts of predictions with respect to both phenomena. I am very dubious about the assertion that we can make ourselves poorer by improving the productive efficiency of a given market.

MR. WRIGHT: Thank you for making that point. I do not think we want to open up another Pandora's box, which is the many and beautiful flaws of the large macro models. The naive model has consistently outperformed these models, which are short-run models, and it has outperformed them in the short run. They do not pass the minimum kinds of credibility tests that we would apply to what those models were designed to do and how they are supposed to be applied and how they try to capture microadjustment. They try, but they regularly fail.

It is hard to deal with Mr. Cooper's comments in a constructive way. I disagree with several points. I think that his evidence that there is no longer a shortage misses the point. He is looking at superficial evidence and inferring things that are not there. The real place to look, as Edward Erickson has pointed out, is in the reserves addition market. The logical extension of Mr. Cooper's argument is that we have made a real mess, it will cost us something to get out of that mess, and so let's not deregulate. That is not a bad paraphrase of some political arguments about why we cannot do it. The longer you leave a policy, a bad policy, in place, the bigger the argument about not doing anything about it.

I want to respond to two points about counting and quantification. I did not say that it is impossible to count; I did say that it is impossible to know with confidence, as we used to. I do not want to leave the impression that it is impossible. I think we can count, but we should recognize that it is going to be game playing. When we look especially at supply elasticity, supply responses, we are in the world of let's pretend, because we do not have a good fix on what those are like. I don't know if we have any way of telling what they are, but we can try to reduce the realm of uncertainty.

CATHERINE ABBOTT: I am always intrigued when Mark Cooper describes the work done by Hudson-Jorgenson and sponsored by my office, because we have such different interpretations of what it says. The significant thing from my point of view about what the analysis says is that yes, there may be short-run adjustment costs in the first three years of phased decontrol. The gains in potential GNP to the economy over time, however, far outweigh—even on a discount basis—those short-run costs. And yes, wealth transfers go on, but the significant thing about the Hudson-Jorgenson analysis is that it is one macroeconomic model that at least attempts to capture and integrate the microeconomic efficiency gains, which all of us who are economists think are so important, into a general dynamic equilibrium kind of framework. It is very significant that the story that emerges from that is exactly the opposite of what Mr. Cooper suggests; that in fact, over time, there are very substantial gains to be had from phasing in full decontrol—on the order of $38 billion on a present-value basis. That is not a small number. That is a substantial counting up of the potential gains to the entire economy. It does not sound to me like trickle up. It sounds like a much bigger pie. It is important to recognize that an attempt has been made to count up both in a partial microeconomic framework and in a macroeconomic framework. Individuals may quibble with pieces of the analysis, but there has been that attempt. We would be better off in the economy as a whole if we addressed this problem sooner rather than later.

BENJAMIN SCHLESINGER: Of all the arguments we have heard about old gas, I don't think supply is one of them. I see no evidence that charging a lot more for old gas is going to enhance the supplies of natural gas in the United States. As Milton Russell's paper pointed out, there would be a lower average price for natural gas under partial decontrol as long as the old gas held out, which would not be very long. Hence, I really am advocating total decontrol when I advocate lagging old gas decontrol. I see the presence of this old gas during the decontrol period of 1985 as enhancing the prospects for the smoothness of decontrol. Whatever effect the contract clauses are going to have, and it could be substantial, that effect will be less if there is old gas in the system, because it will weigh down the average price through the period when the price escalates as a result of contract clauses. Perhaps a small amount of old-gas adjustment is necessary on the well abandonment issue. I do not know how much adjustment is required, but perhaps a supply argument can be made pragmatically on that

155

basis. But I do not see the supply argument at all. I do not see any more gas coming about as a result of some quick action of the FERC on old gas.

GORDON GOOCH: One of the adjusting mechanisms specifically placed in the Natural Gas Policy Act was to adjust the prices of sections 104, 106, and 109 gas, and to adjust them by the old Natural Gas Act's just and reasonable standards. Are we going to write that out of the Act? Are we going to say now that we had our fingers crossed? Under the Natural Gas Act's just and reasonable rate standards, the producers are not earning their just and reasonable rate for their flowing gas supplies. The question is, Are we willing to look at that? It may have the kind of benefits that Arthur Wright has pointed out. But wholly independent of whether it has those benefits, if the rates are not just and reasonable, if the cost-based rates show that gas is being sold below cost, by what mechanism, by what right, do we deny those revenues? If 102, 103, and perhaps other wells are drilled on 104 acreage, even now they can qualify for the 104, 106, or 109 rate—mainly the 104 in this context. So wholly independent of the market-ordering potential, to which I do subscribe, there is also the problem, for those who believe in cost-based rate regulation or believe the FERC should set ceiling prices for gas—let them do it. Let the rates be just and reasonable.

MILTON RUSSELL: We need to keep in mind when talking about moving gas prices up that it makes a very big difference which of those prices move up. The proposal in the paper by Wright and Jacoby was to move forward on section 104 gas. I share many of the concerns about the econometric models— what they capture and what they do not capture. It is important to know which gas prices are moving up, because it makes a great deal of difference to distribution among pipelines and among consumers of different sorts. And it makes, therefore, a great deal of difference to how the gas cushion is dissipated. To suggest that one method of dissipating the gas cushion is as good as another is to suggest that consumers and constituents do not recognize whose ox is being gored under which circumstances.

I have suggested that under partial decontrol the price would probably be lower than under full decontrol. I think that is correct, but let me make sure we recognize that it is correct, ignoring the possibility of a contracts problem that will kick the prices up temporarily for some contracts and at the same time recognizing that we are talking about short-term frictions. In my judgment at least, the way to think about the gas problem after 1985 is to assume that those markets are going to clear at roughly the same prices as under full decontrol.

CHARLES STALON: It appears that the sense of the speakers is that the rents embodied in the old gas will be dissipated sometime soon—1985, 1986, 1987—and I would like to pose this question. What will we leave the state public utility commissions with in 1985 or 1987 once we have dissipated the rents? Are we going to end up with partial deregulation, which means partial regulation, which transfers no rents to ratepayers? Will we end up with prices the equivalent of market-clearing prices, which would come about without any deregulation? We are stuck with the inefficiencies that Milton Russell mentioned. One of those inefficiencies will be the retention of a vintage pricing system, caused partially by the control mechanism and partially by long-term contracts that are not fully indexed. That vintage pricing system will probably be accompanied by a rolled-in pricing system at the city gate, which will leave us with a dysfunctional set of prices. The wellhead price will reflect the marginal social cost of gas, but because of the rolled-in pricing mechanism, the distributing companies will have a price that does not reflect the marginal social cost of gas plus the transportation cost. Any attempt to integrate a synthetic gas industry into the natural gas industry to complement it will tend to be undertaken either by the pipelines or by the fuel producers. Distributing companies will be discouraged from innovating in the synthetic gas industry.

For an industry, or at least a society, that is trying to move to competitive markets to discipline the field price of natural gas, this is a strange set of circumstances. I have never believed that competitive markets arise by immaculate conception. They arise by hard work in a society creating them. The consequence of all this seems to be that the state commissions are going to be confronted with terrible pressures to exploit every demand curve that they can find to keep some of the distributing companies afloat. I believe the consensus around this table is projecting a very unsatisfactory set of circumstances into 1985 and 1986 for state commissions. It seems to boil down to one of two things. They can fight to restore some of those economic rents by stretching out regulation. I really don't think that is possible; I think Mr. Leone covered that very well. Those rents will be squandered. So I would argue that that solution is not an appropriate one. The only rational solution we have is to proceed with deregulation, but this time with the explicit purpose, uppermost in our minds, that the objective is to create a competitive market for gas. It is not to get the government off the back of the producers. That does not end the problem.

My fear as I listen to people around the table is that by 1985 we will loosen the Gordian knot and things will work out—we will have a market price that clears. Who will be left to fight the battles of all the inefficiencies that are still hanging around? A few academic economists waving marginal cost curves. Where is all the industry going to be? They will have given up. They will have got the government off their back, but we will have created a dysfunctional market rather than a competitive and highly functional market.

MR. STALON: No, I am not arguing for decontrol now. I am arguing, however, that by 1985, when we really have dissipated all the rents, some of which are going to ratepayers, we will have a constructively dysfunctional market. I am willing to continue to take some of the rents for a while and, for reasons of political survivability, phase in this whole problem. So I am sympathetic to a proposal to begin to escalate some of the old prices in order to reach a condition in 1985 that is politically viable and has long-term promise. I could not afford, in my current position, to advocate and tolerate the consequences of immediate decontrol.

MR. WRIGHT: I'm not sure I understand what you would want in 1985 that would be better than what we might have under any of the options. Unless it was to take care of all of the side issues—which I don't mean to say are unimportant—like contracts. That certainly will be around to haunt regulators. You are not advocating getting public utility commissions off the back of the local distributors?

MR. STALON: No, not at all, for a lot of reasons. There is a natural monopoly still. My point is merely that this whole debate seems still to be motivated, directed, and powered by the interests of the industry. And their particular positions, many of them with great merit, will I believe be satisfied one way or another by 1985, 1986, or 1987. But we will still be left with a set of dysfunctional prices in the system and no benefits to the ratepayer.

GRANT THOMPSON: I have two comments, one on the paper and one on something Mark Cooper said that I would like to dispute. It seems to me that the solution the authors are trying to present to us is one that inevitably opens up Pandora's box. In trying to deal either administratively or through congressional action with just a few categories of gas—since that is where the money is and what the fight is about—we will get a court battle if FERC tries to do it or, if FERC does it and wins the court battle, a windfall tax. I sense a certain nervousness that Wright and Jacoby's preferred solution helps us very much to keep the lid on Pandora's box shut. I worry whether we really need not think through some of the harder issues that option four tries to finesse.

The second point is that I heard Mark Cooper dismiss rather lightly the improvements in efficiency on the demand side that could come from decontrol. I certainly do not want to let that go unchallenged. Coming from the Conservation Foundation, how else could I approach this? In other energy forms we have seen large efficiency improvements. We have seen consumers in industries make readjustments and efficiency adjustments, not simply fuel-switching adjustments, that have surprised all the experts and sent them back to their models to change the coefficients. One could expect the same kinds of good surprises as a result of decontrol. We could see a great loss from the R & D advances that price can induce in the efficient use of natural gas, particularly as the federal government gets out of the conservation research business. This is yet another reason to let prices rise so that we pull forward those efficiency improvements on the demand side.

JACK EARNEST: I think Mr. Thompson covered one of the major points I wanted to make; that is, if there are efficiencies to be achieved with total decontrol, playing with the old gas categories and taking care of the so-called cushion between now and 1985 or 1986, which somebody may benefit from, might not be a complete answer. Mr. Thompson has indicated further that he believes we need to get total decontrol to get the benefits. This is something we ought to focus on.

I am not sure we can do it by dealing only with old gas because, depending on how the contract problem is handled, I can see totally inequitable charges against consumers that we just cannot stand. That comes about through the contract problem and through the higher prices, which have been subsidized to some degree by the old gas. To the degree that happens between now and 1985 or 1986, a problem is created that has to be solved—prices that, for a large part of the gas supply are far above a market-clearing level. Somebody—typically residential and commercial users—will have to absorb those extra costs. I hate to call them inequitable costs, but they are costs in excess of the price at which gas ought to clear. Some industrial customers may be able to avoid them, which would create further problems for the residential or commercial customer and cause great market distortion.

That is not a very effective way to reach the objective of total decontrol. There must be a better way to get there, and we must come up with some sort of mechanism that will give us the right transition. I don't think playing with old gas is enough to get us where we are trying to go. There are no really perfect answers. Nobody can come up with a system that will solve all the problems. Admittedly, it will have to be a blunt instrument, but it is too blunt to go for total decontrol immediately because the cost is hard to recover. It is too blunt to play only with old gas because that puts a load particularly on residential and commercial customers that they should not have to stand during a transitional process created by thirty-odd years of messing it up.

EDWARD ERICKSON: I feel a little bit like that character in the old television western—have marginal cost curve, will play around. I was particularly interested in Mr. Stalon's arguments about inefficiencies beyond the ones that Arthur Wright and I would focus on. I did not see closure at the end when he asked for response, because I was not sure what he had in mind. When he and I were at a conference where there was a lot of concern, I was amazed at the degree of follow-on questions with regard to making pipelines common carriers. Mr. Stalon, is that something like what you had in mind with regard to parallel reform?

MR. STALON: No, I had in mind primarily seeking complete and full deregulation at some point down the road as well as perhaps some constraints on long-term contracts to keep the vintage pricing system and the rolled-in problem down to an acceptable form. But why haven't we talked about some very fundamental reforms that will get us around some of these problems? Even complete decontrol will leave us with the potential of long-term contracts and with certain regulatory and institutional defects. Common carrier systems for the pipelines at least would be a reform that would get at some very fundamental problems. I am not supporting it; I have some doubts about it. I think it works primarily for the benefit of the big distributing companies. The little distributing companies need those skills of the pipeline.

ROBERT WOODY: I feel obliged to observe, particularly in the company of a member of the conference committee, that we are today experiencing the highest levels of oil and gas drilling in the history of the industry. And it is not all in the Persian basin. As a result we have a 2-trillion-cubic-foot surplus of natural gas in this country today. Gas is so abundant domestically and internationally that many of the problems we now think severe may not in fact be so. I'm from Missouri when it comes to the price spike; I don't think anybody can really prove that. There is some backing off from the dire predictions of six months ago. I also think the presumed problems of the intrastate pipelines are not as severe as we thought they were going to be. In December the Energy Information Administration said there was a sixteen-cent differential. The contracts problem I just do not believe is severe, given that the market is in surplus and supplies are going to be even greater, although the statistical lag is such that I cannot prove that today.

So if we cannot make the argument that we must have higher prices for old gas because we need them for an incentive to drill, we have to use the disorder argument. The cushion is the problem, and I am sitting here thinking, Was the Congress so foolish? Isn't there some advantage in the price-averaging scheme, which permits the search for gas where it is, geologically, and that is in near-deep, tight, and deep formations? Some 75 percent of the gas yet to be found in this country is in those areas. Doesn't it make some sense to encourage and permit the search for that gas without visiting the full consequences on the consumers? It seems to me that makes some sense; so why is that the problem?

If we could play god with the NGPA, we would repeal the Fuel Use Act; we would repeal incremental pricing; we would open up our markets. If we really believe in natural gas, that is what we ought to do. We believe we could, essentially by the end of this decade, back out all of the imported liquids—except for the transportation market and for those cases that may have some foreign policy implications—all that OPEC supplies to us, as strange as that may sound. We think that capability is within our grasp. There may not, just may not, be a knot. But there will be a windfall profits tax, and that will be a form of regulation more permanent and perhaps more severe and more complicated than what we now experience.

MR. WRIGHT: Let me put in a very quiet but determined plug for the position that we took in the paper and clarify a couple of points. I cannot resist the temptation to point out again that the search for section 107 gas was not a deliberate act on the part of the Congress in order to find deep gas. It had some other origin. It is exactly the kind of rent dissipation we do not want. We do not want those rents going down deep holes at this point We want them going down any hole where the cost payout is right. So this is a perfect example of the wrong kind of rent dissipation and a graphic illustration of why we need to do something.

If we need to do something about the Gordian knot, not ignore it and punch deep, deep holes in the earth, what do we do? Nobody claims that focusing only on section 104 gas solves everything. We offered it as no panacea. We did not say not to worry about the contracts problem. I have not heard a good solution to the contracts problem. All I know is that it exists and needs attention, but I do not know how to solve it. Looking at 104 gas focuses our attention on why we are here, which is that something is going to happen out there. There is a big incentive to do the wrong things. The most meritorious thing available is to go after 104 gas and do some other things about contracts and some of the problems that public utility commissions will face. So I offer that in very quiet defense of the position that we state in the paper.

MS. ABBOTT: I cannot resist responding to Mr. Woody's point. The disparate prices on average between the interstate and intrastate markets are not the point. The point is that the decontrol provisions in 1985 are very different. Some old intrastate gas gets decontrolled in 1985. That is the key to the problem, not that there are predominantly different average prices in the interstate and intrastate markets. Indeed, as Robert Means said, there are more substantial differences within the interstate market today than between the markets. The 1985 decontrol provisions are the key to the market disorder problem.

Moreover, if gas is as abundant as you believe it is—and it may well be—that simply suggests that pouring resources at $8 to $10 per thousand cubic feet into deep wells today is an even greater misallocation of resources than we have said it is. Simply to argue that the gas market is going to clear is to say it will clear at lower prices. To say that it will clear at lower prices is to say that the resource misallocation is worse than we said it is. Finally, you have argued that the drilling incentives in the NGPA are working. Why is it that we have had fairly substantial decreases from last year to this year in drilling between 7,500 and 15,000 feet? Those decreases are so big that there is a net decrease in drilling below 7,500 feet, according to American Petroleum Institute statistics, which are public. Yes, there has been a great increase in drilling below 15,000 feet, but the decrease in shallow wells and intermediate depths has more than offset that increase.

MR. ZYCHER: The latest drilling data that I have seen for January 1982 suggest that oil drilling is up 55 percent since January 1981 and gas drilling is down 1 percent. There is a problem in lumping oil and gas together. I am a very firm believer in market processes and always amazed at the ingenuities of the private sector participants, specifically the ingenuity with which arguments can be formulated around distorted price signals to make them sound as if they are sound social policy. Surely one cannot believe, given the history of U.S. Geological Survey estimates of oil and gas reserves, that the U.S. government is a better estimator of where gas is to be found than the private sector would be given the right incentives, which is to say undistorted incentives. I find it a bit questionable to argue that somehow the government, through this process of huddling in smoke-filled rooms for ten months, has miraculously come up with a set of incentives that not only are socially efficient but also make some business firms better off than they might otherwise be.

CHUCK LINDERMAN, Edison Electric Institute: In responding to Mr. Woody and Mr. Wright about the suggestion that the Fuel Use Act be repealed in its entirety, I submit that there will be an efficiency loss for the nation as a whole. For electric utilities that are burning oil and would like to convert to coal, the provision that they can get a direct conversion order from the secretary of energy will be eliminated. That will be a significant loss for those companies and for the nation as a whole.

I would like to have seen a little further amplification in the paper on the role of gas as it is used for electric generation, particularly in the gas-producing states. It was traditionally the cheapest, most cost-effective fuel for the utility industry. Our industry has already made its decision to move wholeheartedly to coal, to lignite, and to nuclear power in the gas-producing states. In Texas, for example, between now and 1990, over 4,000 megawatts of coal, lignite, and nuclear-powered electricity will be brought on line. Somebody might say "That will greatly reduce the demand for gas." But unfortunately the utilities in Texas can take out only some 200-odd megawatts of gas-fired electric generation capacity because of the growth in that state. We will have to continue to burn gas until enough coal and other alternative fuels are brought on line. It may not be true that immediate decontrol or any kind of decontrol will automatically bring more efficiency in the electric generation market because those companies are now and have been moving as fast as their financial capability will permit to some other kinds of fuel.

REPRESENTATIVE PHILIP SHARP: I spent some ten months at a green table in the

House Administration Committee room with a variety of players on the NGPA as we informally met as a conference committee in secret to give you something to discuss today. Had all of you been a part of that conference, we would have had unanimity and rationality and would not have needed these further discussions. Fortunately we decided to give up some of our reformist attitudes about meeting in public, and that did shorten some of the speeches. We found a system by which someone would simply indicate speech one or speech thirty-five and everybody would understand what he was talking about.

I had the additional pleasure of explaining the NGPA to my colleagues in the House of Representatives, who generously supported it. There has been some concern raised here about whether anybody understood it when they voted on it. I will simply keep that a secret. I do want to make clear to you that I supported everything in the NGPA that you consider rational and workable. I opposed all those idiocies that got into it, and some of my colleagues in that conference have said the same. Victory has many fathers, and defeat is always an orphan.

One has to remember the context in which we make some of these decisions. The issue was of great controversy and great complexity that we helped to simplify; but the reality is that we had a political deadlock that nobody had been able to break for twenty years. As a very minor member of the committee the first two years I was in Congress, I saw the deadlock continue although an effort was made to break it. Then it took us two years, with a ten-month negotiation between the House and the Senate finally to break it. Whatever deficiencies and disabilities there are in the NGPA, that was a very significant step for American energy policy and a very important political decision. So I will defend my support and my vote on the basis that it was high time the country made some decisions that took us forward. Although it obviously generated additional problems as well as solved some, it got us beyond what was clearly a deteriorating situation in the natural gas area. I think that situation has been at least marginally improved by passage of the act.

We appreciate your attention to the issue today, and undoubtedly you can appreicate a little more, from those of us who had to deal with the issue in an electoral context, the limited enthusiasm with which people enter into this subject and why you have not seen people on Capitol Hill racing to try to resolve these problems. The overwhelming reason that undoubtedly was addressed by some of you here today is coming to a consensus on what is the compelling reason to act and whether that compelling reason will be the same next year or the year after. It was facetiously said to me that the way world oil prices are going, we may be discussing the fly-down problem next year instead of the fly-up problem. That has almost been the usual history of policy considerations of energy in this country. I trust it is not likely to be the case, but I hope that you will see those of us involved taking seriously some of the things you people had to say. Thank you very much.

163

SELECTED AEI PUBLICATIONS